Sexuality and Medicine
in the
Middle Ages

Sexuality and Medicine
in the
Middle Ages

DANIELLE JACQUART
and
CLAUDE THOMASSET

Translated by Matthew Adamson

Princeton University Press
Princeton, New Jersey

Library of Congress Cataloging-in-Publication Data

Jacquart, Danielle.
   Sexuality and medicine in the Middle Ages.

   Translation of: Sexualité et savoir médical au
Moyen Age.
   Bibliography: p.
   Includes index.
   1. Medicine, Medieval. 2. Sex—History. I. Thomasset,
Claude Alexandre. II. Title.
R141.J3313 1988   306.7′09′02   88–25311
ISBN 0–691–05550–5 (U.S.)

Contents

Preface

We would like to express our gratitude and our acknowledgements to our friends and masters, Guy Beaujouan and Mirko D. Grmek. Both of them, through their teaching and their constant kindness, have given us encouragement, help and support. We would like to thank the translator of the English edition, who has managed to find, wherever possible, texts in Middle English corresponding to the medieval French versions we quoted. We have been able not only to include, in the English edition, illustrations and an index, but also to add extra details and to up-date some of the bibliographical information.

Danielle Jacquart and Claude Thomasset

Introduction

Anyone who wants to learn more about the way people lived in the Middle Ages has to turn to those representations of the world that they left us. Their scientific texts reveal, without any admixture of poetic fiction, the order of a universe which, they believed, obeyed the will of God and in which the Christian could read God's design. But this entire world was there to serve men and women; their bodies carried out those functions necessary to the survival and continuity of the species. The fleshly exterior was ephemeral, bound to disappear as it awaited resurrection, and at first theologians would not accept that a scholar could devote time and effort to the body, which was seen as of only secondary importance: the churchmen wanted first and foremost to be doctors of souls. Having for several centuries been a domain reserved for monks, the medical treatment of the body little by little became autonomous; granted full recognition in the thirteenth century, which was when medieval science made its real conquests, medicine moved only gradually into the hospitals, which long remained devoted principally to the care of souls. Theologians' attitudes did not, however, hinder the continuous development of a corpus of medical and scientific knowledge which, if one leaves aside popular medicine, grew concurrently with the rediscovery and extension of ancient knowledge. The exploration of the laws of nature became legitimate only slowly: the unravelling of the world's secrets and of the mysteries of the human organism was often denounced as a Promethean endeavour, lying outside the legitimate field of investigation into the human creature.

It should not be deduced from the reserve maintained by theologians, or from their over-assertive doctrinal statements, that people in the high Middle Ages had no desire to use scientific knowledge to avoid illness. The survival of vestiges of ancient, and more particularly Hippocratic, medicine shows the care that people took in handing on medical knowledge – and this happened long before Isidore of Seville, undertaking the task of collecting and summarizing the philosophical and scientific learning that had been preserved in spite of the fall of the Roman Empire, created the first monument of medieval knowledge. This determination to explain the world, to understand the signs that God had laid down in things, was always accompanied by a more pragmatic approach, in which the theoretical system of explanation was applied to the remedying of dysfunctions in the human body. Thus diets, prescriptions and medical advice forced their way into theory and were recorded in the medieval encyclopaedias. On the other hand, people tried to fit a whole body of medical practice, perhaps based on observation, into the rigid classifications, the simple and logical models with which the Middle Ages tried to grasp the way the living body functioned. This, in brief, is the evidence we shall be examining.

The subject of our enquiry is medieval sexuality, and our choice of this subject needs to be justified. The function of reproduction was at the centre of theologians' and doctors' concerns. Gynaecology and embryology have been dealt with in numerous studies, and so we will not be taking our investigations any further than the exact moment the embryo is first formed. In dealing with a mentality which could not imagine that a woman might be anything other than continuously fertile, and which refused – at least in theory – to assign to the sexual act any goal other than procreation, it may seem a rather dubious procedure to try to describe the ways sexuality was represented while leaving out of the picture gynaecology and the constraints that this one goal of procreation necessarily imposed. Nevertheless, an enquiry into sexuality is often a study of the means of avoiding conception and will bring to the fore those authors who treat pleasure as primary. In other words, our discussion must grant the greatest importance to that which is implicit – to the allusive elements of discourse. Considerations and techniques relating to all forms of sexual life developed in curious ways, alongside a discourse that tended to repress them, whether in the shape of priests imposing penances or that of doctors wielding arguments of a physiological nature.

In our quest for signs of change, we must thus isolate the detail, grant pride of place to the phrase or the word that shows a new way of thinking in a piece of writing which, taken as a whole, seems to be highly conformist. Evaluating the degree of active support or passive acceptance shown by whoever made the text available to the public is also a matter of the greatest interest. A certain section of a work giving details of some strange practice seems to have awakened no echo in the mind of its commentator. To be more precise, certain passages in Avicenna – and this is true of other texts – seem to have been ignored or misunderstood, while others were no sooner read than they gave rise to intellectual speculation. Furthermore, one and the same person could speak in different voices: a doctor of the Church, for instance, when undertaking scientific research or indulging in the extraordinary intellectual competition of commenting on a text by Aristotle, could come out with opinions that might cause problems for his theological thinking. Finally, in the very heart of a text of uncertain provenance, some blunt interpolation may provide evidence of contemporary thought. Numerous texts of scientific popularization, though ill-defined in general outline, are the source of valuable information. Even more than the diversity of voices within a work or a text, language itself has to be seen as both a vehicle of information and as an obstacle to the proper evaluation of the knowledge of sexuality.

Erotic pleasure stems in part from the stimulation of the imagination, and Arabic literature, for example, which combined poetry and didacticism, had shown a clear awareness of this function of reading. Thus the collecting of texts – whatever their scientific content – had played and continued to play this role. The accusation of self-indulgence sprang immediately to the lips of detractors. William of Conches had to insist on the legitimacy of his discussions of sexuality and gynaecology. The censor was already lying in wait and the author had to anticipate him by claiming the right to self-expression. The pleasure of taking up the challenge could not be ignored and in this way a dialogue was established between the two parties concerned: it was to last for centuries, with all the psychological complications inherent in the subject. But we will say no more about the problems peculiar to discussions of sexuality.

More simply, as in every branch of medical knowledge – and perhaps more than in any other – a certain opacity of vocabulary came into play. The traditional difficulty in the naming of organs was a constant problem, with all the dangers involved in successive translations – from Greek to Arabic, from Arabic to Latin, from Latin to Old French. The possibilities of mistranslation were thereby multiplied and in the course of history wrong interpretations gave rise to real scientific errors. A new translation was always an important event from the point of view of science, for a number of concepts needed to be defined more precisely. At the final stage of transmission, when the word had to be expressed in Old French, the difficulty was even greater, since scientific thought had to cast itself in the mould of a language which did not possess a vocabulary suited to it. Borrowings from Greek and Latin allowed one to get round the difficulty, and the resources of the vernacular were pressed into service – a particular word being given a new meaning in the process. Scientific language and everyday language thus coexisted, so that within each word a common and a technical meaning were brought together under a single signifier. In the sphere of sexuality, the term of scholarly origin and the everyday term were found side by side, as if it were necessary to make a constant pedagogic effort for those who did not have sufficient knowledge of the Greek and, especially, Latin etymologies. It has of course to be admitted that the gap which nowadays separates the two levels of language is even wider than in the Middle Ages. Those who express themselves in the language doctors use are granted the right to talk about sexuality; those who use everyday language are accused of indulging in vulgarity.

These, however, are surmountable obstacles compared with that created by the formidable coding used by a particular social group to talk about sexual practices. In a given situation, any word or phrase may contain a sexual allusion and any text may give rise to a plurality of meanings. This may sometimes lead to extravagant interpretations – all the more easily since, in any given synchronic whole, every kind of language has the power to suggest an analogy, to inspire a metaphor, or to create an allusion. After a few years, this second network, superimposed on the text's immediate meaning, becomes inaccessible, or at least partly so. While the most elementary metaphors are relatively easy to detect because of their permanence, those meanings which denote types of sexual behaviour are difficult to decipher. Lack of understanding of the associative network is a crippling handicap when it comes to the discourse of people as experienced in the art of double meaning as were scholars; in addition, the Latin language encouraged certain types of verbal juggling. Furthermore, a bilingual situation prevailed, in which the most *risqué* puns could easily cross from one language to another so as to be hidden from the eyes of the profane, while being immediately revealed to initiates. It seems that the scholars of the Middle Ages had nothing to learn when it came to the art of alluding to the most delicate, indeed the most censored subjects. In short, the erotic art displayed so explicitly in other civilizations may have circulated by virtue of the semi-transparency of a language – or of two different languages – capable of conveying a plurality of meanings.

Nonetheless, our desire to decipher these words and to establish a collection of terms enabling us to explore the depiction of sexual activity can be only one of the intentions that justify this piece of research. The system of representations of reality was an equilibrium that underwent imperceptible modifications, and each modification, however small it might be, was proof of transformations that could concern other fields of knowledge, for medicine and all the other sciences were interdependent.

The need to understand the mechanisms of the physiological system was only a part of the desire to bring the complexity of reality under control and, in accordance with Aristotelian procedure, physics as well as the mechanical arts contributed, by virtue of analogy, their explanatory models to it. In fact, the object of our enquiry is not only the progress of scientific knowledge but also the functioning of the system of causality particular to a civilization and grasped as one synchronic whole. Rather than thinking of sexuality and embryology as an application of the principles of other sciences, we should on the contrary imagine erotic and embryological knowledge as capable of forming a model for other areas of scientific and spiritual research. We are here thinking of alchemy.

Thus, when the medieval doctor attempted to elucidate the mysteries of the living organism, he transgressed the domains established by modern science. But this assimilative and all-conquering discourse borrowed not only from past and present knowledge, but also, when it judged it useful, appropriated the inventions of theology and the legacy of popular thought. Anatomical or physiological explanation was a melting-pot in which scientific knowledge was amalgamated with the *exempla* that had been inherited from a long tradition. Superstitious beliefs became more important than causality, when it could no longer be perceived – or, more simply, such beliefs replaced causality altogether, whenever the scientific observer yielded to the pressure of his or her cultural heritage. On this score, medieval science reveals surprising mental blocks as well as astonishing intuitions.

Perhaps more important than scientific progress, which was often a mere accompaniment to the rediscovery of the texts of antiquity, was the freedom acquired by discourse. Control of erotic experience meant discovering the possibilities of the body, giving a new dimension to time and trying to escape from the narrow determinism imposed by a too exclusively teleological vision of humanity's role on earth. Rejecting the instinctive aim of the sexual act while still claiming the right to pleasure was a more difficult enterprise than it seems. The most highly educated sector of the population made every effort to find a place for itself in society and, with this in view, made use of the scientific technique and knowledge of sexuality. Scholars took a leading role in this, thanks to their discovery of the function of play. The games that bodies played and the techniques of love could be expressed through the ludic and erotic possibilities of language. It would be wrong, however, to criticize them for going about this in a light-hearted way.

The Middle Ages were, after all, heavy with menace: the erotic game, and games with eroticism, can be explained as a way of keeping fear at bay and resisting the diminution of human freedom. Sexual activity was linked to fear in the medieval West – and not only the fear inspired by moral or religious law, the transgressing of which filled men and women with the terror of the sacred. On a more prosaic level, anyone who reads the medical texts with care can discover at every step, lying in wait, the obsessions of medieval people; any impairing of their organisms could arouse fear – whether it was fatigue, physical disability, or growing old. The most dangerous moment was when man encountered woman: so many moralists had warned him of the dangers he was running! Woman was the creature of discontinuous time; she was a threat during menstruation, often out of bounds during pregnancy and the nursing period and prohibited on holy days. The union of man and woman became almost an adventure. Doctors' efforts consisted not in overcoming superstition – such a task was impossible – but in attempting to explain the physiological and psychological specificity of woman. The greatest minds accomplished marvels in this sphere, at a time when the most virulent attacks on women were being made. But however attentive medieval medicine might be to the body and even to the mind, it remained, in the majority of serious cases, powerless to cure. Even more, it was completely defenceless when faced with the greatest scourges. Every attempt to produce a rational explanation of epidemic transmission remained fruitless. Hence there continued to exist, in doctors' writings and in popular awareness, irrational stories and mythical histories that gave an immediate solution or irrefutable proof in the form of the tale or the *exemplum*. Yet illness did find a place in medieval thought. One need only think of leprosy, but venereal diseases should be remembered too – mutilating bodies, developing out of control and giving rise to the gravest anxieties. These diseases aggravated men's lack of understanding, indeed their hatred, of women. Should one be surprised if, in conditions such as these, scientific writing remined susceptible to the pressures of theology, and was exposed to the influence of foreign cultures at their most irrational?

Medieval scientific thought relating to sexuality was all-pervasive. It could be found in literature and in the foreign literatures from which it took ideas accepted as emphatic truth; in the diversity of Arabic translations, from which it increasingly drew sustenance; and in the theological treatises that were responsible for assigning to men and women their places in the world. These questions were present in the thinking of all intellectuals: some of them wanted to codify, while others wanted to find out and to cure. The former and their extravagances have often been discussed; it is only fair that the voice of the others should be heard too.

1

Anatomy, or the Quest for Words

The medieval anatomist is most often presented, these days, as a theoretician, blinded by the confidence he placed in the authority of Galen, whose authenticity he merely sought to verify. This vision, which was to a large extent handed down to us by Renaissance scholars determined to break away from a culture that no longer satisfied them, does not take into account the conditions necessary for an observation to have any effect. The success of any observation depends on the underlying knowledge that will enable meaning to be given to what is seen. In the field of anatomy, it is particularly difficult to follow a detailed description, when it is unaccompanied by any pictorial representation and especially, as was the case in the Middle Ages, when the texts which present the description have been translated successively from one language into another. Thus the aim of the earliest dissections, carried out at first on animals and then on human corpses, was to help one to understand the original sources, an understanding that could not come about through textual interpretation alone. The contribution made by the Middle Ages was to render Galen's discoveries intelligible again and to impose order on them.

Medieval Galenism was not, however, the same as authentic Galenism; Owsei Temkin has shown how the first deviations came about within the Alexandrian school.[1] The divergence increased over the centuries through exposure to other influences, which were added to earlier ideas without eradicating them. At the end of this development, the fifteenth century was marked by the trace of each successive influx that had filled out the gaps in the piecemeal knowledge handed down from low Latin antiquity. This characteristic of medieval thought, which superimposed rather than selecting, was particularly evident in anatomy, a science less in thrall to intellectual fashions and philosophical currents than was physiology. The persistence of successive influences explains in particular the unstable and heterogeneous character of the vocabulary: synonyms proliferate even within the work of a single author.

The description of the reproductive organs provides us with an especially good example of the magnitude of the divergences that can be found between one author and another. To the modern way of thinking, which has inherited from the nineteenth century the idea of organic function, this instability of anatomical data seems irreconcilable with a coherent physiological system. From the point of view of medieval physiology, based on the mechanism of humours and forces transmitted by the spirits (or *pneuma*), organs were considered merely as channels, receptacles or vehicles. Their existence was explained in terms of their purpose, and not of any specific function they might perform.

WORD-PLAY

Etymological exercises crop up constantly in the field of anatomical description. This peculiarity of medieval thought explains the interest we shall be taking in it, as we embark on our discussion of the way the organs of the human body were described. Indeed, Isidore of Seville, in the sixth and seventh centuries, created, in his *Etymologiae*, a form of knowledge that the Middle Ages were to repudiate only much later.

Jacques Fontaine has shown clearly how a certain type of scientific discourse was set up, and how, by means of language, power was exercised over the way things were represented.[2] The 'grammarians' were skilled enough to establish an intellectual approach founded on the analysis or splitting of the signifier, so as to make this the essential activity underlying all knowledge. It has been demonstrated that this form of thought belongs at once to the pagan Hellenistic inheritance and to a Semitic tradition transmitted by the Bible. According to Jacques Fontaine, the role of etymology was simultaneously 'novel and essential' in the transitional culture of that period, at a time when the intellectual no longer possessed anything more than the vestiges of ancient culture, while medieval culture had not yet been born.[3] The word, and the etymological exercises based on it, constituted both the first elements of that new culture and at the same time a constraint, an obligation which it was very difficult to evade. It seemed inconceivable that one might question the guide-lines proposed by Isidore, for the knowledge that he was sensed to have behind him, as well as his intellectual standing, made him the first authority. He was indeed the first from a purely chronological standpoint, but also the foremost representative of this particular way of thinking, since he supplied the words and set out guide-lines for the network of analogies based on them.

It is perhaps interesting to recall the way Isidore's etymologies work.

Right from the start, we can exclude from consideration those cases in which the imposing of a name is the expression of a human activity, 'according to convention' (*secundum placitum*): an examination of the form of such words can clearly reveal nothing essential.[4] Other cases, however, may reveal to us the hidden face of the particular creature or thing under consideration. The definition of etymology proposed by Isidore, debated so frequently, is the following: 'ETYMOLOGIA est origo vocabulorum, cum vis verbi vel nominis per interpretationem colligitur.' Joseph Engels translates: 'The origin of (the reason for) ways of naming is an etymology, when the meaning of the verb or noun is grasped by means of an interpretation.' Isidore's etymology, in its most complete form, is a complex analysis which can be set out in the following manner, taking as an example *nepos* (descendant, grandson = born after the son):

Nepos	=	Nomen, vocabulum
Qui ex filio natus est	=	Vis nominis
Quasi natus post	=	Nota
Primum enim filius nascitur, deinde nepos	=	Origo-veriloquium[5]

What is most interesting to the modern mind is the *nota*, that is, the 'vocal kernel' around which the name was formed, this *nota* being stated expressly or preceded by *quasi*. Here is an example taken from the description of the human body, which presents the head as the starting-point of the nerves, and thus as the origin of all strength: 'Prima pars corporis CAPUT; datumque illi hoc nomen eo quod sensus omnes et nervi inde initium CAPIANT, atque ex eo omnis vigendi causa oriatur.' The principal interest of etymology for our purpose is that it enables the essential function of the organ under consideration to be grasped.

We will not leave Isidore of Seville without reading how he gives an even more precise definition of his working instrument:

The reason for ways of naming is an *ETYMOLOGIA* (i.e. *VERILOQUIUM*), that is to say something in conformity with the nature of things, when the *NOTA RERUM* fashions nouns and verbs, and as a consequence the *VIS* of the noun or verb can be grasped by means of an *INTERPRETATIO (NOTATIO* or *ARGUMENTUM EX NOTA)*, that is to say by means of an *ETYMOLOGIA*.[7]

Even more than this method of investigating the name, which we have just sketched out rapidly, Isidore bequeathed to the Middle Ages a procedure which, in his work, leads to a 'religious revelation'.[8] So the real motivation for medieval philological research was to be the urge to discover the fragment of hidden truth concealed in each linguistic sign.

The Middle Ages adopted the concept of etymology rather loosely defined by Isidore. Without going into the details of the different interpretations and controversies, it will be a good idea, so as to show the interest that twelfth-century grammarians took in etymology, to mention a few names and a few books. One must cite Hugh of Saint-Victor and his treatise *De Grammatica*, and especially Pierre Elie, who in the *Summa Prisciani* proposed a twofold definition of etymology which was frequently quoted and commented on; for him, etymology

– either discovers the 'primam vocabuli originem',
– or is an 'expositio alicuius vocabuli per aliud vocabulum'.[9]

In the second hypothesis the concept of *etymologia* is included in that of *expositio*. The word submitted to etymological investigation has to be explained by one or more other words better known to the reader and which fulfil the conditions proper to etymological activity – which basically involves accounting for the word's phonetic structure and the properties of the particular creature or thing under consideration. The question of the origin of the name is no longer relevant. Etymology must satisfy extralinguistic criteria, and aim to describe the essence of the referent.

Despite the efforts of certain grammarians to establish distinctions between the different concepts, the Middle Ages seem to have accepted, under the heading 'Etymology', Isidore's definition in its widest sense.[10] Roswitha Klinck shows that the thinking of grammarians and even of theoreticians of poetry led to the establishing of a system of medieval etymology that was far removed from the demands of the grammarians of ancient times.[11] The new doctrine allowed one to examine the word practically without limit. The slightest analogy between two terms could become the starting-point of an etymology. This could be adapted to the meaning of the text in which the word occurred, and thus become an argument used by the commentator. Several etymologies for one and the same word could coexist. At this stage, etymology allowed every flight of fantasy and was no longer an impediment to thought. Too systematic an adoption of Isidores 'dictionary' had managed to exercise an over-decisive influence on people's thinking; the very abuses of the practice restored to thinking a freedom which certain medieval authors did not fail to exploit to excess.

The space we have devoted to Isidore and his conception of language shows how important he was for medieval civilization. Etymology laid bare, with all the dryness of a logical demonstration, the very essence of medieval language. In his work, as in that of all writers in the centuries after him, language's every sign was linked to its referent by a series of correspondences that had to be deciphered. The signifier could be split up in many ways and each of its parts revealed an aspect of the particular creature or thing signified. Michel Foucault showed that this absolute, sacred motivation of the sign was at work in the sixteenth century too:

In its raw, historical sixteenth-century being, language is not an arbitrary system; it has been set down in the world and forms a part of it, both because things themselves hide and manifest their own enigma like a language and because words offer themselves to men as things to be deciphered. The great metaphor of the book that one opens, that one pores over and reads in order to know nature, is merely the reverse and visible side of another transference, and a much deeper one, which forces language to reside in the world, among the plants, the herbs, the stones, and the animals.[12]

Anyone venturing into the domain of medieval scientific thought would do well to remember this.

Isidore's medical knowledge is gathered in books IV and XI of the *Etymologiae*: book IV discusses diseases and their cures, while book XI is devoted to anatomical description. This knowledge is meant for the laity, and for clerics who wish to find out a specific piece of information, while not themselves necessarily being either specialists or practitioners.

Following the traditions of Greek thought, Isidore's universe is made up of four elements. Each element possesses two qualities: relations can be established and transformations occur between any two elements that share a common quality. Hippocratic medicine had established the correspondence between the elements and the humours of the human body (blood, phlegm, black bile and yellow bile). The state of the human body was determined by an equilibrium between the humours (*temperamentum*). Thus was established the principle of a correspondence between microcosm (the human being) and macrocosm (the universe), an idea that can be found in Isidore's thought.

In the section devoted to anatomy, the etymologies show a desire to exalt the benefits of God's wisdom. The way each organ, with its own characteristic qualities, corresponds adequately to its final purpose, reveals a harmonious physiological conception of human life, in which the teleological impulse is never found wanting. The anatomical description surveys the human body from head to foot, but we will be considering only those organs that have to do with sexuality.

It is our own, modern attitude to questions of sexuality that leads us to choose the passage on the female breasts. Nothing in the work of later encyclopaedists could enable one to consider the female breasts as being endowed with any particular erogenous sensitivity; only their function as a source of food was indicated. Here is Isidore's text:

Mamillae vocatae, quia rotundae sunt quasi malae, per diminutionem scilicet. Papillae capita mamillarum sunt, quas sugentes comprehendunt. Et dictae papillae, quod eas infantes quasi pappant, dum lac sugunt. Proinde mamilla est omnis eminentia uberis, papilla vero breve illud unde lac trahitur. Ubera dicta, vel quia uberta, vel quia uvida, humore scilicet lactis in more uvarum plena. Lac vim nominis colore trahit, quod sit albus liquor leucos enim Graece album dicunt: cuius natura ex sanguine commutatur.[13]

We are, of course, quoting the text in Latin so that the reader will be able to see the way it plays on etymologies. They may be classified in accordance with the different categories that Isidore himself proposed. In the first group there are the etymologies that are coined *ex causa, ex origine* or *ex contrariis*; in the second appear those that are produced *ex nominum derivatione, ex vocibus* and *ex graeca etymologia orta*.[14] The other possible classificatory system is suggested by the tradition of Augustinian rhetoric used by medieval authors, the two principal axes of which are the *translatio similitudinis* and the *translatio vicinitatis*.[15] The relation 'apple-breast' is a good *translatio similitudinis*, which is then developed by a process of word-formation – in this case, a diminutive suffix is added. The sequence *ubera* – breast, via *ubera lacte* – in which there is abundant milk, via *uvida* – full of liquid, juicy, in order to arrive at *uva* (the breast is full of milk just as the grape is full of juice) constitutes an extraordinary *tour de force*. One has to confess that while philology and science lose out somewhat, poetry reigns supreme. It would be wrong to smile at this example, for the etymology of the word *lac* draws in reality on the Galenic theory of dealbation, in which blood is transformed into milk – a theory that all authors of the Middle Ages fully believed in and which was frequently discussed in great detail.

Isidore's constant eagerness for etymological research very often led him to take into account the function of the organ under consideration. This tradition, as Jacques Fontaine has shown, was deeply rooted: 'The preoccupation with functional explanation was linked, in Latin medicine, to a Stoic conception of the role of purpose in nature.'[16] The fundamental ideas of pagan philosophy were thus at the origin of the teleological principles implied by the acceptance of the role of Providence in nature. Thus in Isidore one finds a methodical quest for purpose. With the help of etymology, Isidore's biological finalism could be harnessed to an art of apologetics. In conjunction with this biological teleology came a functional teleology, which consisted for example in pointing out protective functions (the protection of the organ of sight by the cheek-bones, eyelashes, eyebrows and eyelids). Bearing in mind that aesthetic teleology can be found in the *Etymologiae*, we may turn our attention to a fourth form of teleology – a rather unfortunate creation on the author's part, since it cuts across the categories already mentioned; the 'distinctive teleology' of organs created or differentiated 'so as to act as distinguishing marks'; such is the function of the beard in enabling one to distinguish between the sexes.[17] Since we now know the rules of the Isidorean game we can quote a few more etymologies.

The kidneys (*renes*) are so named because *rivi ab his obsceni humoris nascuntur*, i.e., the stream (*rivus*) of semen (*obscenus humor*) springs from them.[18] We may note in passing that sperm (*semen*) comes from the spinal cord, according to a theory found in ancient Greek medicine and which we shall be encountering again. The loins (*lumbus*) are the seat of lust (*ob libidinis lasciviam dicti* = 'because of the lewdness of desire'), but in the woman, the seat of lust is the navel (*umbilicus*).[19] The passage presents us with a fine example of the way Isidore's thought works. The principal idea is that the seat of *libido* is in the centre of the body, whence the series of subsequent associations, which we will give in the opposite order from that of the text:

— *umbo* (the swelling in the centre of the shield);
— *umbus iliorum*;
— *umbilicus*;
— *l* + *umbus* (*lumbus*).

It is in this way, we imagine, that the first-named object is joined to the analogical series. The verbal sequence has been created thus: the human body is seen as being arranged in the same way as the world (the microcosm–macrocosm analogy underlies this move), and the requirement that *libido* should be given a central place leads to a quest for signifiers that will demonstrate the correctness of the theory. Isidore handed on to the Middle Ages the idea of this localization of the seat of lust – and so of an erogenous zone – in women; it still held an important place in the encyclopaedists of the thirteenth century, notably in the work of Vincent of Beauvais. This entirely theoretical notion could well contain a grain of truth: Isidore could be merely expressing a male fantasy which sees the female navel as having an especially important role to play.

The *genitalia*[20] are the parts of the body which, as their very name indicates, have been so called because of the way they engender offspring, because it is through them that procreation and begetting are effected. They are also called *pudenda*, shameful parts, because of the shame associated with them, or else because of the hair (*pubes*) which appears on them at the age of puberty. One may also cite the etymology of *veretrum* (penis), which is explained thus: *viri est tantum* (it is proper to the male), or else because the semen (*virus*) is the humour that flows from the sexual organs of the man (*vir*). As for the testicles (*testiculi*), they are the diminutive of the word for witness (*testis*); for there to be testimony, there must be at least two people. Indeed, right at the beginning of the Middle Ages, a language was established that revealed a certain conception of woman; man was already proclaimed as the complete being who held no mystery; and all this was done with a method and a language of formidable efficiency. Man drew his name (*vir*) from his force (*vis*), whereas woman (*mulier*) drew hers from her softness (*mollities*). One may disregard the fact that language, supposedly of divine origin, needed to be deciphered so as to yield the truth concealed in it – a truth that constituted a final argument – since the appeal of etymology could easily be explained by an appeal to more ordinary psychological considerations; its associative and mnemotechnic power was by itself perfectly capable of leaving its imprint in people's minds. Etymological proofs structured not only consciousness, but also the unconscious. Woman was physically weaker so that she could be subjected to man, so that she could not repel his desire – for, once rejected, he might then turn to other objects.[21] Isidore's teleology tried to justify every particular detail in order to establish a harmony of the world that avoided the misogynistic extravagances that were to be committed by certain theologians several centuries later. Nonetheless, man's fear when confronted by woman's insatiable sexual activity is already evident: 'Females are much more sensual (*libidinosiores*) in human beings just as in animals.'[22] Thus Isidore transmitted to his successors, first, a science of medicine that drew its explanatory principles from the elements and their qualities; secondly, the fundamental analogy between microcosm and macrocosm; and thirdly, the Pythagorean theory of the origin of sperm.[23] Above all, he gave them a method that would be all the more successful as it was seen as the heir to the ruins of ancient culture.

We must now ask what fortunes this method was to meet with. In the ninth century Rabanus Maurus, in his ambitious project the *De Universo*, took over without any change a substantial number of Isidore's etymologies. Examples relevant to our topic are the words *matrix*, *vulva* and *semen*. In addition, he added an allegorical interpretation to Isidore's etymology. Thus he glossed the word 'menses' (*menstrua*), from the Greek *mênê* = moon,[24] with the following commentary: 'It is not allowed to approach menstruating women nor to have intercourse with them, because a Catholic man is not allowed to have anything to do with the idolatry of pagans or with the heresy of heretics'.[26]

An etymology – well-founded, as is the case here – followed by an allegory leads us into all the extravagances of the analogical game, which may make it impossible to observe any object forced to undergo this treatment, concealing it behind all the associations of ideas that immediately come to mind. Linked to the moon's pernicious action, menses – and a list of the damage they cause follows – are not only dangerous physically, but through being associated with heresy evoke some of the sacred horror that accompanies the transgression of a taboo. This idea could already be found in the Bible, but this way of putting it, this association of ideas, gave clear proof of its truth. For if the first proposition was true, the second (one must not have any dealings with heretics or idolaters) was even truer and, in consequence, it was now out of the question to cast doubt on the first proposition. The physical world and religious law were linked in such a sophisticated way that it would be naive to deny the persuasive power of such a demonstration.

So much for those dangerous exercises that were to find their place in the works of the most misogynistic theologians; we must now move on to discuss those scholars who made a practical effort to construct a less systematically preconceived body of knowledge.

The etymologies of Isidore of Seville that we have quoted crop up again in the works of encyclopaedists of the thirteenth century; in the *Speculum Naturale* of Vincent of Beauvais or in the *De Proprietatibus Rerum* of Bartholomew the Englishman, 'Ut dicit Isidorus' appears at the head of every article devoted to this or that organ. These etymologies are strangely absent from the *De Natura Rerum* of Thomas Cantimpré, who refers only to Galen and Aristotle. Etymology can however still be found in the anatomical descriptions of the masters of Salerno. This is true of the word 'vulva', which Isidore discusses in these terms: 'The *vulva* is named by analogy to a folding door, *valva*, that is, the door of the belly, because it receives the semen, or because the foetus proceeds from it.'[26] In the *Anatomia Magistri Nicolai Physici* one finds the following addition: 'It is called *vulva* from *volo vis* or else from *volvendo* [*volvere* = to roll an object, to roll something into shape].'[27] This example shows clearly the great flexibility with which the teleological principle can be adapted; the function of the vulva is to mix the two seeds, and the movement transmitted is at the very origin of life; one here finds the reminiscence of an idea found in Hippocrates and a fine example of the way medieval thought worked. This example enables one to understand the long survival of etymology; it continued to be productive because it was capable of giving rise to several analogies, and thus of supplying arguments for different types of explanations. Even if etymology could not abandon its first apologetic vocation, it was allowed, en route, to make its own contribution to scientific explanation, since the sometimes difficult reconciling of theology with science was part of the role played by the religious authorities.

Though medieval thought rid itself of the acrobatic exercises of

Augustinian moralizing at the beginning of the fourteenth century, etymology was far more tenacious. An interesting example is supplied by the examination of a medieval encyclopaedia, written around 1320, the *Compendium Philosophiae*.[28] While allegory no longer had any place in it, etymology was all-pervasive and the compiler was so steeped in it that he no longer felt it necessary to cite his author. The etymologies of Isidore long kept their evocative power; in his otherwise audacious commentary on the *Anatomia* of Mondino de' Luzzi, Berengario da Carpi was still, in 1521, recording the meaning of *vulva* and *umbilicus*.[29]

Our few pages devoted to Isidore have enabled us to show that philology in the service of theology played a very important part in medieval thought. Etymology was merely the consequence of a certain conception of language. The question is even more complex if one takes into consideration those works written in the vernacular, for there was a hierarchy between different languages and three of them were sacred: Hebrew, Greek and Latin. A Latin etymology in a text written in Old French possessed an enhanced value as argument since it was a step in the direction of the original language. The same was true, *a fortiori*, for a Greek etymology appearing in the middle of a Latin text. Anybody embarking on a study of the sciences in the Middle Ages must necessarily take the precaution of dismissing the idea that there was any break between language as it pleased God to give it to humanity on the one hand, and scientific language on the other. Language always made it possible to invent an argument that confirmed the way a doctor imagined an organ. In its materiality it constituted an inexhaustible reservoir of signs.

Nonetheless the progressive disappearance of explanation by etymology could not be halted; the disfavour it fell into was in direct proportion to the degree to which the work of Aristotle became known. The *Cratylus* had instanced various reservations about this way of thinking, and Aristotle insisted on the arbitrary value of the sign in language.

Another phenomenon doubtless played a determining role; the increasingly technical nature of books devoted to anatomy and, concurrently, the spread of a terminology of Arabic origin, still rather restricted in the Salernitan period, but making increasing inroads thanks to the numerous transliterations made by Gerard of Cremona. Such terms constituted a highly technical vocabulary that hamstrung the imagination of commentators incapable of indulging in word-play on 'barbarian' words that were more tiresome than evocative – all the more so since they came from a language which itself excluded etymologies and at the same time was incapable of revealing any truth to Christians. Faced with the influx of these neologisms, authors regularly preferred to turn for help to the *Hermeneumata* or the *Synonymi*, kinds of bilingual lexicons which set out a Greek or Arabic term on one side and, opposite it, a paraphrase meant to stand in for an as yet non-existent Latin word; no extralinguistic etymological activity was possible at this level. Establishing a scientific language was not without its difficulties. A learned commentator of the fifteenth century, Jacques Despars, showed a certain perplexity when faced with different ways of designating what today is called the great omentum: 'This part of the body is called *tirbum* by Avicenna, or, according to other manuscripts, *zirbum adipinum*; *omentum* by Hippocrates; *epiploon* by Galen, in the *De Utilitate Particularum*; *adomen* in the *De Complexionibus*; *mappa ventris* by the new surgeons; and *zirbus* in common language.'[30] An examination of these different words shows, in Latin writing of the later Middle Ages, the persistence of an Isidorean 'reflex'. The Greek word was based on the form and location of the organ; Oribasius noted: 'It is claimed that the *epiploon* received this name because it floats (ἐπιπλέω), so to speak, on the intestines.'[31] The Arabic word *tharb*, transcribed as *zirbum* or *tirbum*, designated the fatty substance it is composed of, expressed by the Latin adjective *adipinus*. As for the expression adopted by Western surgeons, *mappa ventris* (a covering or clothing for the stomach), it was an attempt to express the organ's function – its purpose.

Iconographical depictions did not belie this tendency. Apart from the figures illustrating the different positions of the foetus and which came from the *Gynaecia* of Moschion, manuscripts only exceptionally included an iconography of our topic.[32] Particular attention must be paid to a series of illustrations which, although corresponding to the state of pre-Salernitan medicine, continued to be reproduced even after anatomical description had been improved as a result of Arab influence. These illustrations thus contributed to maintaining a certain archaism. Two of these figures showed the male and female organs. In no sense did they constitute a realistic representation, but were, rather, diagrams whose purpose was to explain physiological processes. If we take the example of the diagram that represents the female organs, we see that one of the manuscripts shows particularly clearly the modifications undergone by the uterus during pregnancy;[33] the two states of the woman are reproduced in one and the same figure, which makes the diagram extremely 'theoretical'. This type of representation was so far from being realistic that another manuscript stuck the captions proper to the female sex onto a barely distorted diagram of the male apparatus. This error was no accident; as we will show, the Middle Ages adopted the ancient idea of an inverse similarity of the genital organs of men and women.

The same diagram could thus be used to show the male organs or the uterus, which was bicornate according to medieval descriptions.

As in the other fields of anatomy, the spreading practice of dissection hardly modified the way the sexual organs were represented. Certain of the miniatures that accompany the *Anatomy* of Guy of Vigevano (fourteenth century), however, show the traces of 'lived experience' – as

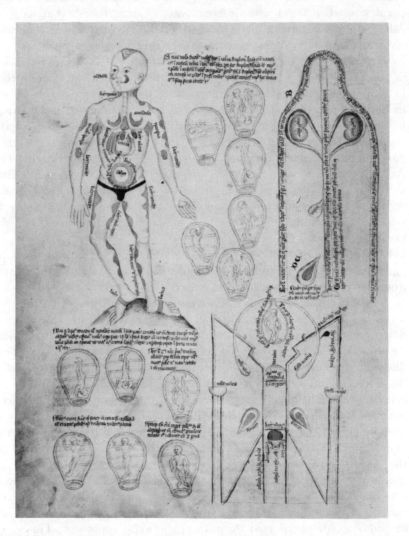

FIGURE 1.1   *A representation of female anatomy showing also the different positions of the foetus in the womb.* Apocalypsis, *Wellcome Ms 49. fol. 37v. (Wellcome Institute Library, London)*

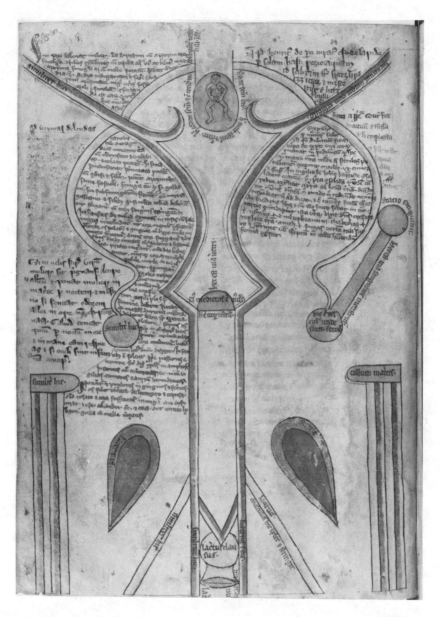

FIGURE 1.2 *A thirteen-century drawing of the anatomy of the uterus and adnexa. Bodleian, Ms Ashmole 399. fol. 13v. (Bodleian Library, Oxford)*

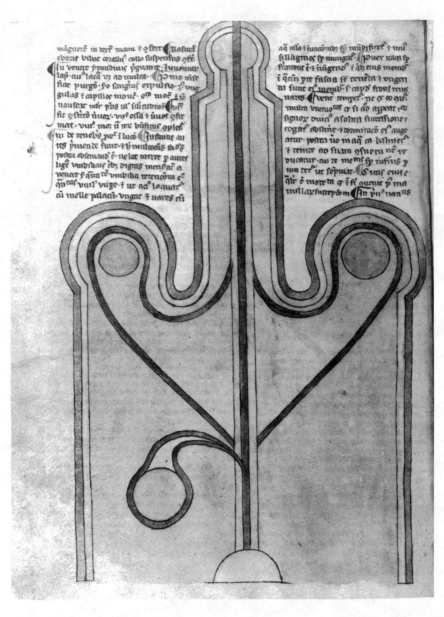

FIGURE 1.3 *A thirteenth-century representation of the male genitalia. Bodleian, Ms Ashmole 399. fol. 24v. (Bodleian Library, Oxford)*

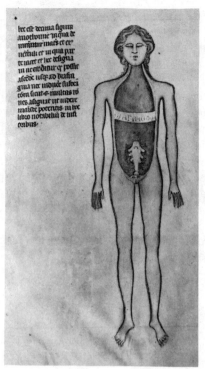

FIGURE 1.4   *The tenth figure of the treatise of Guy of Vigevano, showing a woman whose womb has seven compartments. Musée Condé, Chantilly, Ms 334/569. (Photo. Lauros-Giraudon)*

in the series representing the different layers of the abdomen.[34] The same cannot be said of the female body drawn in one of the miniatures; if the grey colouring of the corpse was supposed to suggest a will to realism, the shape of the uterus is in brutal contrast. We will quote Ernest Wickersheimer's description:

The uterus is represented as Mondino described it, that is to say, it is comprised of seven cells, three on each side and one in the middle. Ligaments shaped like fish-hooks start from the uterus and radiate out in every direction, but none of them reaches the diaphragm. In this figure, the author's purpose was above all to demonstrate that the uterus cannot, as certain authors believed, reach as far as the diaphragm and there cause the disturbances known under the name of suffocation of the womb.[35]

Medieval anatomy continued to be explanatory rather than descriptive.

THE FREE PLAY OF ANALOGY

The decisive impetus to scientific development was given, in anatomy as in several other areas, at the end of the eleventh century, by the *Pantegni*, a translation of the *Al-Kunnās al-malikī* of 'Alī ibn al-'Abbās al-Majūsī. The author of the Latin translation, Constantine the African, has over the centuries gained the reputation of being a skilful forger; it is true that he omits to cite the name of the Arab author and gives a Greek title to what he presents as a compilation of the best ancient sources. It has, however, to be granted that while he takes the liberty of cutting out passages in an often arbitrary and inappropriate manner, Constantine translates literally the passages that he does quote. (For a chronology of the main translations from Arabic into Greek, see table 1.1).

We can judge the quality of this, the first transmission of Arabic culture, by comparing, with the help of short extracts, the original texts and Constantine's translation of the *Pantegni*. 'Alī ibn al-'Abbās begins

TABLE 1.1 *Chronology of the main translations from Arabic into Greek*

Century	Translator or adaptator	Works translated
6th	Moschion	*Gynaecia* of Soranus of Ephesus (1st c.)
11th	Alfanus of Salerno	*De Natura Hominis* of Nemesius of Emesa (4th c.)
11th	Constantine the African	*Pantegni* of 'Alī ibn al-'Abbās (10th c.) *Viaticum* of ibn al-Jazzār (10th c.) *De Coitu* of ibn al-Jazzār (?) *De Spermate* (pseudo-Galen)
12th	Gerard of Cremona	*Canon* of Avicenna (11th c.) *Liber ad Almansorem* of Rhazes (9th–10th c.) *Chirurgia* of Albucasis (10th–11th c.)
13th	Michael Scot	*De Animalibus* of Aristotle (4th c. BC) *De Animalibus* of Avicenna
13th	Bonacosa	*Colliget* of Averroes (12th c.)
13th	William of Moerbeke	*De Animalibus* of Aristotle
13th–14th	John of Capua	*Theisir* of Avenzoar (12th c.) *De Coitu of Maimonides* (12th c.)
14th	Nicholas of Reggio	*De Usu Partium* of Galen (2nd c.)

his chapter on the description of the genital organs with a brief introduction which Constantine does not translate:

Having said enough about the digestive organs, it is necessary to speak here of the way the parts of the body called the genital organs are arranged. These parts of the body are: the womb, the breasts, the testicles, the spermatic ducts and the penis. We will begin first of all with the womb, and we will discuss its shape, its position, its uses, and the conditions of the foetus in this organ.[36]

The description of the womb was translated faithfully by Constantine, but the decisions that he had to make in order to find an adequate Latin vocabulary were to have important consequences.

FORM AND SUBSTANCE OF THE WOMB

Al-Kunnās al-Malikī So I say that the womb, and especially its base, is similar in shape to the bladder, but it is different from the bladder in that it possesses two lateral extensions similar to horns, which stretch to the groins; the veins and arteries which bring blood and pneuma to the womb enter via these extensions. In substance the womb is similar to the nerves, because it has to stretch in all directions during pregnancy, while the foetus is growing.[37]

Pantegni The womb (*matrix*) is like the bladder in shape, for both of them are very deep, but it is different in its two extensions which are similar to horns. These horns extend as far as the emunctories and contain arteries which bring blood and spirit to the *vulva*. The substance of the *vulva* is, as it were, sinewy, so that it may stretch in all directions when conception has taken place and when the foetus begins to grow.[38]

THE TUNIC OF THE WOMB

Al-Kunnās al-Malikī The womb has just one tunic made of fibres running in different directions. It has fibres which stretch out lengthwise; these fibres are few in number, since they are only necessary in order to attract the sperm. It also has fibres which stretch out slantwise; these fibres are more numerous because of the force that is necessary to retain the sperm and the foetus during pregnancy. Finally, it has fibres which stretch out breadthwise, because of the expulsive force necessary at the moment the foetus leaves the womb.[39]

Pantegni There is just one membrane (*panniculus*) which includes hairs (*pili*) assembled in different shapes. Some are small, stretch out lengthwise, and are necessary to attract the sperm; others stretch out slantwise, and are bigger, so as to retain the sperm and the foetus once it has been conceived; others, stretching breadthwise, are there to expel the foetus at the moment it leaves the womb.[40]

The 'cervix' and the cavities of the womb

Al-Kunnās al-Malikī   The cervix of the womb [= vagina] extends to the slit, that is, the empty space situated between the two bones of the pubis, and this cervix is next to the anus; on the outside it has extensions of the skin, called clitoris [= labia?]; it is analogous to the prepuce in man and its use consists in guarding and protecting the womb from cold air. The womb has two large cavities, one on the right, the other on the left. These two cavities end in a single cervix shared by both of them, which is called the cervix of the womb.[41]

Pantegni   The cervix is linked to the nature of the woman – the nature of the woman being a little space situated between the two bones of the pubis – and it is next to the anus; on the outside it possesses little flaps of skin called *badedera*, which are analogous to the prepuce in men; they are necessary to protect the *vulva* from cold air. The womb (*matrix*) has two large cavities, one on the right, the other on the left. These two cavities end in a single cervix which is called *collum vulve*.[42]

The principal difficulty encountered by Constantine seems to reside in the absence of a set of linguistic tools in Latin, rather than in ignorance of the Arabic. To translate the single word 'uterus', he hesitates between *matrix* and *vulva*. This ambiguity is not only characteristic of the *Pantegni*; the manuscripts and the first editions of Galen's commentary on the *Aphorisms* of Hippocrates, also translated by Constantine, show the same fluctuation.[43]

Indeed, in the Latin of late antiquity, the word *vulva* designated a rather vague semantic field; it tended, depending on the authors, to designate either the woman's external genital apparatus taken as a whole, or, in the work of certain writers, more specifically the womb. Constantine the African uses the old terminology at the same time as he introduces a new one. To designate the vulva specifically, he makes use of the expression 'the nature of the woman', taken from Cicero.[44] The Arabic text is not particularly clear either, when it is no longer describing the uterus; while the word *bazr* generally refers to the clitoris, the explanation seems more suited to the labia minora. Here the translator was merely transliterating the Arabic in the form *badedera* – which was to lead future authors to ignore those parts of the female sexual apparatus, all the more since they play no role in conception, properly speaking. However, the Latin West had had at its disposal since the sixth century Soranus's precise description, which he had taken from Moschion:

What is the woman's sinus? A nervous membrane similar to that of the large intestine: very spacious on the inside, it is, on the outside – where coitus and the acts of love take place – rather narrow; it is vulgarly called *cunnus*; outside are the labia, called *pterigomata* in Greek and *pinnacula* in Latin; from the upper part there comes down into the middle what is called *landica*.[45]

That this information – which was, after all, easily available – fell into oblivion was no accident: for the medieval mentality, the important thing was to explain the mechanism of reproduction, not to describe in any precision the appearance of the organs. So it is not astonishing that, in anatomical accounts, there persisted statements that we know today to be false: their success or failure depended on the role they played in the general explanatory process. Thus the Middle Ages continued to believe in the existence of cavities and villosities on the inside of the womb. We will have several occasions to return to the former: how many of them there are was to be one of the principal causes for disagreement between authors. As far as the villosities are concerned, we must point out a particular feature of Constantine's translation. The Arabic text uses the word *layf* which, like the Greek word *inê*, designates any fibre of vegetable origin, to describe the kinds of roughness of surface that are favourable to movements within the womb. Its translation into Latin as 'hair', in Salernitan and post-Salernitan medicine, was to meet with a success that was all the greater as the dissection of the sow confirmed it: in this animal, the uterus is indeed particularly villous,[46] quite the opposite of what can be observed in woman.

In the *Dragmaticon Philosophiae* of William of Conches, the 'villosities' are dealt with in special detail. He says: 'Prostitutes after frequent acts of coitus have their womb clogged with dirt (*oblimatam*) and the villosities in which the semen should be retained are covered over; that is why, like greased marble, the womb immediately rejects what it receives.' If one reads through the full anatomical description given by William of Conches, one notices another hollow internal organ covered with 'villosities' – the stomach, which is thus equipped so as to be more able to retain food.[47] In his description of the womb, William of Conches contrasted villosities, capacity for retention and fecundity on the one hand with marble and everything that is smooth and sterile on the other hand. It seems that an error in translation lay behind this anatomical notion. Words in all their opacity do not enable us to reconstitute the real way the medieval doctor interpreted reality. The most plausible explanation lies in analogical thinking: from real 'villosities' – those in the intestines, for instance – it was easy to deduce the existence of 'villosities' in the womb. Thought created a representation of the organ that was perfectly adapted to showing its Galenic 'retentive' virtue. The absorption of food, and embryonic growth, both occurred amidst these subtle, burgeoning threads, which managed to establish an exchange between matter and the living organism. In this example, as in others that we will be encountering, it seems that the pre-scientific imagination was dominated by the idea of the process of nutrition in the tree. The villosities are an image of the roots of a plant which draws its life from inert matter. The burgeoning of capillaries is the metaphor meant to explain the appearance or survival of a living being.

The theoretical knowledge transmitted by Constantine found an application in those exercises of anatomical description, accompanying the dissection of a pig, which were in fact the first Salernitan anatomies. Needless to say, the practice of dissecting a corpse did not exist. Indeed, it seems that Galen himself had never dissected anything other than monkeys, and could thus be criticized for applying to the human body observations that had been carried out on animals. In Salerno, dissections of the pig were carried out – in conformity with the precept that one meets in all the anatomies: on the outside, the human body resembles a monkey (or a bear); on the inside, a pig.

From the Salernitan anatomies as studied by George W. Corner,[48] we will mention three major texts that will serve as guides. The first two were probably written beween 1100 and 1150, the last at some indefinite date in the second half of the twelfth century, though certainly before the Toledan translations were in circulation.

1 *Anatomia Cophonis* or *Anatomia Porci*. G.W. Corner hesitates whether to recognize under this title a revised pre-Constantinian work, or a post-Constantinian work written by an author who did not grasp the vocabulary properly. It can thus be taken as an example of the difficult transition from one level of knowledge to another.[49]

2 *Second Salernitan Demonstration*. This work is critical of what is said in the above-mentioned *Anatomia*, but it is likewise designed to accompany the public dissection of a pig.[50]

3 *Anatomia Magistri Nicolai Physici*. With a similar incipit, placing the work under the authority of Galen, there exist three versions of the same text, in different forms. The first is attributed to Ricardus Salernitanus, the second is sometimes presented as part of the *Micrologus* of Ricardus Anglicus, the third and most extensive goes under the name of the magister Nicolaus Physicus. These texts all go beyond the restrictive order imposed by the dissection of a pig. Their remarks apply more directly to the human body and the anatomical demonstration, especially in the version of Nicolaus Physicus, shows a sure grasp of the sources available before the Toledan period.[51]

Description of the female genital organs in the Anatomia Cophonis

The sow is laid out on her back, and dissection begins at the throat; working progressively downwards, the anatomist reaches the uterus:

It is next necessary to discuss the anatomy of the uterus [*matrix*]. It must be recognized that nature has contrived this organ in women in order that whatever superfluities are generated during the course of the month may be sent to this organ as if to form the bilge-water of the whole body; this is the nature of the menses which women have. This organ is also nature's field, which is cultivated that it may bear fruit; in which, when seed is sown, it remains as on good ground and through the cooperative action of natural warmth, and the mediation of vital spirits, it becomes implanted like a germinating seed, and sends out twigs through certain roots or mouths by which it is attached to the uterus, and through which nutriment is delivered to it and to the future foetus. Thus, later on, by the action of the bodily forces (as I have often told you, you may recall) the foetus-to-be is generated and augmented. The uterus is located above the intestine; above its neck [= vagina] is the bladder, and under it the *longaon*. Below is the vulva. Next cut the uterus through the middle of its os [cervix]; you will find two testicles [= ovaries] attached above it, by which the female seed is transmitted to the uterus and joins the male seed to form the foetus. The uterus has seven cells, and if the animal is pregnant, you will find the foetuses in these chambers. Over them you will find a kind of tunic, like a chemise, which is called *secundine*. This is broken when the foetus strives for exit. It is attached to the uterus and to the foetus by veins which run in it, and it carries nutriment to the uterus and to the foetus. Those openings by which the foetus is attached are called cotyledons.[52]

This text is extremely revealing. One is struck by the mixture of practical details and theoretical explanations which give it its magisterial style. As for the content, we will leave to one side the discussion of the existence of female sperm which we will be dealing with in our chapter on physiology. In this text one finds precise medical statement united with explanatory analogy. Biblical comparison is followed by the image of the embryo represented as a plant or, rather, as an upside-down tree, a depiction which the West took from Constantine. It can also be observed that the analogy is misunderstood, which does not add to the text's clarity. In any case, it was an attempt to help one to understand the process, for it was impossible to name the thing, the matter (the union of the two sperms) that was to become a living creature: it seems that there was a mental block here, that imagination failed when faced with something that was a mystery for medieval people. It must be said that the absence of any real scientific explanation was further confused by the perplexity caused by the moral and theological question. The

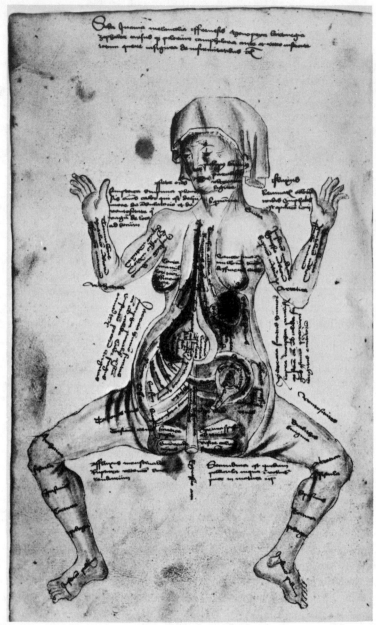

FIGURE 1.5 *A representation of female anatomy showing the reproductive organs. Bayer. Staatsbibliothek Cod. Germ. 597, fol. 259v. (Reproduced by kind permission of Bayer. Staatsbibliothek München)*

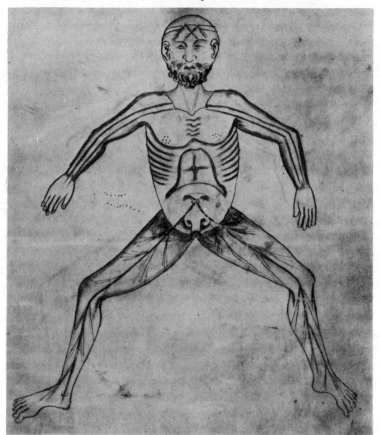

FIGURE 1.6 *The male genital organs. Öffentliche Biblothek der Universität Basel D.II.11, fol. 171r. (Reproduced by kind permission of Universitäts Bibliothek Basel)*

Church granted life to the embryo only on the fourth day of its existence: there was a sort of fear underlying this refusal to name the stage of development which belonged to matter, and was thus part of the realm of physics, since there was not yet even the most rudimentary resemblance to a human being.

Another peculiarity of this text which, we repeat, is an account of the anatomy of the sow, is that it claims there are cells in the uterus. It is thus evidence for the resurgence of an ancient tradition, the success of which was not to be diminished even when the *Pantegni* was in circulation. We will examine it in connection with another text of Salernitan stock, which reflects it in greater detail.

Description of the male and female genital apparatus in the
Second Salernitan Demonstration

While it too is a commentary on the dissection of a pig, the *Second Salernitan Demonstration* reproduces faithfully the assertions made in the new translations from Arabic to Latin, especially in the *Pantegni*, whole passages from which recur in it. The solid constituent parts of the body, indiscriminately called members, whether they are tissues or organs, are classified in accordance with the tripartite division of the faculties which govern physiology. A distinction is drawn between the animal members, the agents of sensation and movement; the spiritual members, which ensure the circulation of the vital spirit; and the natural members, which include the agents of nutrition and reproduction. This latter category has as principal members the testicles, the seat of the faculty of reproduction; to the testicles are joined organs of protection, for example the scrotum, and organs of purification, such as the seminal ducts; there are, finally, parts of the apparatus considered to be instrumental or subordinate: the womb and the penis. Before describing the genital organs in this order of classification, the text refers to the way they are supplied with blood and *pneuma*: the great artery which leads down from the heart along the vertebral column, the aorta (*adorthi*), divides at the level of the kidneys and the vital spirit is distributed along its ramifications to the testicles; in the same way, certain veins, ill-defined in the text, but which, one imagines, are those that issue from the inferior vena cava, branch down on each side of the vertebral column to provide blood.

The testicles, which are the instruments of the sperm, are formed of glandular, white, soft, and spongy flesh, in order that sperm may be generated in them. Each is covered by a membrane, which is derived from the *siphac* (peritoneum). The substance of the sperm before it comes to the testicle is received in a certain follicle, in which it is altered and whitened, and this membrane is below the kidneys and above the testicles; in some animals there is found in the said membrane a great quantity of that moisture [*humiditas*] which is the material of the sperm; in other animals little is found, and in others none; and as we have shown you, there are two passages, one on each side of the membrane, through which this material descends to the testicles. Proceeding from the inferior part of the testicles are two vessels called *seminalia*, through which the sperm passes from the testicles to the penis, and these vessels are long, white, and hard like muscular flesh; they are long, so that the testicular excretion may better undergo coction as it passes along, and broad, that the sperm may pass quickly from these vessels into the penis and from the penis into the female pudenda. In your presence I have incised one of these ducts and have shown you the sperm.[53]

In this passage, the *Second Salernitan Demonstration* faithfully follows the *Pantegni*, in particular in its description of the substance of the testicles and that of the vasa deferentia. The analogy in the text between the substance of the testicles and the matter of which sperm consists recurs constantly in the physiological explanations that we shall be examining. On the other hand, the dealbation undergone by the substance of the sperm, situated by 'Alī ibn al-'Abbās in the circumvolutions of the spermatic vessels, takes place at first, says the Salernitan author, in a 'follicle'. We can interpret this as the vaginal involucrum observable in certain domestic animals, in which it appears as a hernia of the peritoneum, taking the form of a serous sack surrounded by membrane walls. The description could scarcely be applied to the male, since in his case the canal that communicates between the abdominal cavity and the vaginal involucrum disappears after the foetal period.[54]

Unlike the *Pantegni*, the Salernitan text does not give a joint description of the male and female testicles. The animal which it dissects is male, and so it moves on to a description of the penis.

The penis is fleshy, nervous, round, and hollow, beginning at the two *ossa pectines*; and is formed of two cords [?] placed side by side transversely [bulbo- and ischiocavernosus muscles?], which is necessary for double cause. First, that it may eject the sperm into the vulva; for this reason it is made nervous, in order that by virtue of its great sensitiveness there may be intense pleasure even in so unseemly an act as emission of the sperm. It is hollow, in order that in the presence of ardent desire it may be extended and erected with the greatest possible rigidity by means of much spirit in its large cavity and in the muscles placed at its sides; and thus, it is not readily deflected, but may be inserted directly into the vulva. The second necessary cause is that it may expel the urine passing through it from the neck of the bladder without interruption and without harm, as we demonstrated very clearly in the dissection by means of a quill inserted through the neck of the bladder.[55]

Even if the experiment performed with a quill comes at the end to give this description of the penis the tone of a personal example, it is a faithful copy of the *Pantegni*, which itself refers to a Galenic model, especially in the way it explains the mechanism of erection by a pneumatic principle. The paragraph devoted to the female organs on the same page says nothing to deny this influence:

The uterus [...] has two orifices, one external, which is properly called *collum matricis*, in which coitus is completed; the other internal, which is properly called *os matricis*, and this closes, according to Hippocrates, after the seventh hour of conception, and will not thereafter admit the point of a needle. The os itself is nervous and somewhat sensitive, in order that much delectation may be caused in intercourse by contact of this organ with the male member, and it is moderately firm, in order that it may easily be distended for the entrance of the sperm and closed when the sperm is received; for if it were not so – if it were over-hard or over-soft – it would be inextensible through hardness or for softness could not be shut.

We also find once more the assertion (verified *de visu*, if the empirical conditions can be believed) that the womb is covered on the inside with villosities meant to retain the sperm. There are thus only two cavities in the womb, contrary to what we had read before in the *Anatomia Cophonis*. We learn too that the female testicles are smaller and harder than a man's, round, somewhat flattened on the surface and glandular. From the kidneys a vein extends to each of these testicles, and they are situated beneath the horn-shaped extremities of the uterus. From each testicle rises a duct through which the organ ejects the sperm into its spermatic vessel. The discussion of the genital organs concludes with a few remarks on the *collum matricis*, which differs with the woman's age and nature, the stage of the menstrual cycle she is at, and whether or not it has been stretched by childbirth. In young girls and older women, this organ is smaller than in adults and it is wider in ardent women than in the less passionate.

The synthesis of influences in the
Anatomia Magistri Nicolai Physici

While the *Second Salernitan Demonstration* presented the dissection of a pig in the light of the *Pategni*, the last text that we have placed in the same general tendency as that of the School of Salerno opens the way to a more systematic and more theoretical anatomy, at the same time as it shows more diverse influences. The content of the translations from Arabic into Latin is better synthesized with the old underlying Graeco–Latin tradition, and this synthesis encourages constant etymological exercises.

Right from its first line, the *Anatomia Magistri Nicolai Physici* puts itself under the patronage of Galen, but we will see that the use of apocryphal sources fills in any gaps left by the authentic writings. The author recalls the hostility of Catholics towards the dissection of corpses as well as the principle of analogy between human beings and certain animals, then he sets out a classification of the solid parts of the body which is more complete than the list given in the preceding anatomy. This classification begins with a definition of the word *membrum* as every part of an animal which is firm and solid, homogeneous or heterogeneous in composition, and fulfils a specific function. The homogeneous members, called *consimilia*, are the skin, the veins, the arteries, etc. The heterogeneous members or organs, called *officialia*, are formed of a collection of *consimilia* and include four principal members: the brain, the heart, the liver and the testicles, each of which is the seat of a vital faculty. Despite setting out this classification, which had been usual ever since Galen, the *Anatomia Magistri Nicolai Physici* follows in its plan a tripartite division between animal, spiritual and natural members.

In the series of texts in the Salernitan tradition, each anatomical description adds something to the description that had come before. Among the new elements that concern the male genital apparatus, we find a greater precision in the division of functions between membranes, vasa deferentia and sperm ducts:[56]

- the scrotum, a small pouch in which the testicles hang, ensures the protection of these organs; it protects them from the pressure of the thighs and in their turn the thighs protect the testicles from external wounds;
- the lower spermatic vessels [= vasa deferentia] expel waste matter, and carry the sperm to the place of conception through the penis;
- subordinate to the testicles, the 'didymi', or vasa afferentia, transport the raw material out of which the sperm is made; they are called *didymi* or 'dubious', because there is doubt as to their origin (do they rise in the kidneys or the testicles?).[57]

While underlining the importance of this new etymology which makes 'double' the same as 'doubt', we notice above all what a good example the *Anatomia Magistri Nicolai Physici* is of Galenism as the Middle Ages understood it. The care Galen took in specifying the purpose of each part of the body fitted in perfectly with the medieval ambition of demonstrating in everything the perfection of God's intention.

All writers insisted on the extreme sensitivity of the genital organs, described as *nervous*. Here is a description of the place occupied by the genitals in the general scheme of the nervous system as presented by our text:

Two nerves arise at the end of the dorsal spine, one of which passes to the right leg and the other to the left; they descend the whole length of the thighs and lower legs into the feet and there divide according to the number of the toes. Through these nerves animal spirits are borne to produce motion and sensation. Other nerves arise from the spinal cord and pass through the spine to reach the chest and the perineum. The union of these nerves forms the penis in the male sex, and therefore this organ is extremely sensitive and has even been called by some *cauda nervorum*. In the female sex a similar concourse of nerves forms the foramen, or gateway, or vagina.[58]

Several peculiarities concerning the movement of the blood should also be noted. In particular, the existence of the *vene iuveniles*: the author lists the different possible origins of sperm. For Hippocrates, it is formed in the brain, for Galen, in the liver; it originates in the whole body, say others. Hippocrates's opinion is based on the existence of the *juvenile veins*, located on each side of the head: once they have been severed, the man can no longer produce sperm.[59] Unlike men, women have a vein called *kiveris vena*, which Nicholas glosses as *female vein*. Its function, if we may summarize the discussion we shall be devoting to it in our analysis of the physiological mechanisms, is to bring part of the menstrual blood into the womb, and to bring the other part of this blood into the mammary glands so that it can there be transformed into blood, in the course of gestation. This vein probably corresponds to that branch of the vena cava which, called *vena quinaria* in the *Pantegni*,[60] ensures communication between the breasts and the uterus. Under the name of 'quilin', it was likewise described in an ancient commentary on the *Aphorisms* of Hippocrates sometimes attributed to Oribasius. The proposition 'Should the breasts of a woman with child suddenly become thin, she miscarries' is glossed thus: 'The pregnant woman has a vein coming out of the liver, called the "*quilin*". It divides into two: one branch carries the blood to the breasts, and because of its new location transforms it into milk; the other branch goes to the womb.'[61]

It is on this primitive description of the blood's irrigation system that the anatomy of Nicholas is based. He describes the womb in accordance with tradition: this organ is cold and dry in temperament (whereas the testicles are hot and dry), equipped with villosities on the inside, and divided into seven cells. The etymology of the word vulva quoted above also appears, as well as the use of the seven cells to determine the sex of the embryo: 'The womb is divided into seven cells, three on the right, three on the left, and the seventh in the middle.' The author notes the opinion according to which males are begotten in the right-hand cells and females in the left-hand, while the central cell is reserved for hermaphrodites.

One can here see the direct influence of a pseudo-Galenic treatise, translated from Greek into Latin and used from the twelfth century onwards: the *De Spermate*.[62] This text, which contains a particularly sophisticated account of the determinant influence of heredity, linked with astrology, is the principal vehicle of the theory of the seven cells. This theory evidently belongs to the Pythagorean tradition of numerological speculation; the importance of the number seven was emphasized in Macrobius's commentary to the *Dream of Scipio*, which enjoyed a particular vogue in the twelfth century.[63] The role played by the seventh day in conception and by the seventh month in gestation, as they are defined by Hippocrates, is there particularly emphasized, as is the way all parts of the body are divided into sevens. Macrobius's influence was probably the determining factor which led certain authors to adopt the theory of the seven cells proposed by the *De Spermate*. This explains the fact that the authentic Galenic description of the two cavities, although in wide public circulation, like the *Pantegni*, was not always accepted.

The Salernitan period was of great importance in the later development of the anatomy of the sexual organs. Although they were presented as a description, the texts could not transcend their teleological point of view. To the reliability of their information, they added the evidence of observation which was to guarantee their value in the eyes of future doctors. Despite the belief of Salernitan anatomists, the pig is not so very similar to the human being. Because of this, a number of beliefs were so deeply rooted and tenacious that the first dissections of human bodies did not succeed in demonstrating their falsity. If one refers to a contemporary description of the uterus of a sow, one can only be struck by the fact that one does come across two horns, villosities and projections that could indeed account for a separation into cells.[64] The demands for a teleological explanation and *de visu* confirmation were indissociable: the cells of the uterus helped in understanding the way sex and character were determined, and the villosities were part of the general process which explained the movement of the fluids, as applied for instance to the digestive organs. It must be pointed out that one thing is missing from Salernitan anatomics: any description of the external female organs.

REALITY INTRUDES

Despite the introduction, in Northern Italy, of the first dissections carried out on human corpses at the end of the thirteenth century and the beginning of the fourteenth, anatomy at the end of the Middle Ages was characterized by a jumble of facts from different places and different periods, which scholars were hard pressed to harmonize into an orderly synthesis. So it is difficult to talk of a linear progress in knowledge; innovations were inseparable from resurgences of archaic ideas whose archaism was not realized as such.

Thanks to Gerard of Cremona and the group of Toledan translators, doctors had at their disposal the essential Arabic texts; Avicenna's *Canon* became the principal reference work, just as the *Pantegni* had been in the Salernitan period. The rediscovery of Galen's anatomical summa, the *De Usu Partium*, thanks to the translation of Nicholas of Reggio in the first half of the fourteenth century, merely added more details; it was properly assimilated only at the end of the fifteenth century and was to form the basis for a discussion and rethinking of the system only in the Renaissance.

The rather chaotic nature of anatomical knowledge at the end of the Middle Ages is shown in the language by the proliferation of synonyms which we have had occasion to mention. This situation was, however, not entirely negative; the increasing accuracy of descriptions, the comparison of frequently contradictory sources with observation and surgical practice, which was developing at the same time, were to bring to light doubts and gaps in the tradition.

First use of the Toledan translations and the end of the Salernitan tradition: the Anatomia Vivorum

The Arabic texts transmitted by Toledan translations at the end of the twelfth century took over forcefully the Aristotelian and Galenic idea of an inverse similarity between the male and female organs. This is clearly expressed in Avicenna's *Canon*:

I say that the instrument of reproduction in the woman is the womb (*matrix*) and that it was created similar to the instrument of reproduction in the man, that is to say the penis and what goes with it. However, one of these instruments is complete and stretches outwards, whereas the other is smaller and held on the inside, to some extent constituting the opposite of the male instrument. The covering of the womb is like the scrotum, the cervix [= vagina] like the penis. There are two testicles in women as in men, but in men they are larger, turned outwards and tend to be spherical in shape; in women they are small, of a rather flattened roundness, and they are located on the inside, in the vulva.[65]

The parallel is drawn on every level of the genital apparatus – sperm ducts, protective membranes, sustentacular muscles; the female attributes are all considered, from this point of view, as smaller in size or fewer in number than their male equivalents. The analogy between the organs of the two sexes was henceforth to be systematic in medieval medical literature. Although a close relation was established between man and woman, the fact remains that woman was described with reference to man. The model, the positive pole of the comparison, was taken as the norm, and the other pole, which was given a negative value, was considered imperfect. In reality, there was in this analogy a correct observation which the strategy of discourse was to transform into an idea that was ultimately dangerous for women. The smaller size of feminine testicles would constitute, for instance, a sort of anatomical proof of the lesser fertility of female 'sperm'. Doctors endeavoured to restrict the role of these organs to that which modern science attributes to Bartholin's glands, as is shown by the remark of the famous anatomist Mondino de' Luzzi, whom we will have occasion to meet again: '[The woman's testicles] are not really testicles like those of the man; they are, rather, like those of the hare, created for the above-mentioned purpose [?] and to create a humidity similar to saliva which is the cause of female pleasure.'[66]

If we return to the analysis of texts in the Salernitan tradition whose development we have been following, we come across a work whose origin is obscure but which, although it comes from the same tradition, uses the Toledan translations. The work in question is the *Anatomia Vivorum*, sometimes ascribed to Galen.[67] Its success was ensured above all by the fact that Thomas of Cantimpré quotes it twenty-two times in the *De Natura Rerum*, introducing it with the words 'Galen says'.[68] In it, the similarity between male and female organs is clearly indicated:

God created the womb to be the instrument and place of reproduction in the woman; the cervix of the womb [= vagina] can be compared to the penis, and its membrane, that is, the envelope of the interior cavity, is like the *oscheum*, that is the scrotum. One can compare the relation which exists between the instrument of reproduction in the man and the instrument of reproduction in the woman to the relation which exists between the seal which leaves its imprint and the impression of the seal in the wax. The woman's instrument has an inverted structure, fixed on the inside, whereas the man's instrument has an inverted structure extending outwards.[69]

What interests the author of the *Anatomia Vivorum* in this parallel is the opposition interior–exterior, rather than any difference in size. Woman is a being held in on the inside. The reference to the seal and the wax echoes the *Dragmaticon Philosophiae* in which William of Conches compared the cells of the womb to the moulds used by those who cast coins.[70] This analogy, applied to anatomy in the *Anatomia Vivorum*, was often used to explain the physiological mechanisms of conception: it was a restatement of the Aristotelian contrast between form and matter.

Sustained by new influences, the author of the *Anatomia Vivorum* tried to work out a more detailed description of the cervix and the vagina than he could find in the texts that were his model:

The cervix of the womb is like a pouch: it is made of little flaps and folds of skin, so that, like the cervix of the bladder [urethra?] it can expand and contract as necessary. When the mouth of the womb contracts, before the moment of child-birth, along its folds in the manner of a rose whose petals have been previously open, the opening of the cervix is so narrow that only urine can pass through it. . . . For the mouth of the womb is double: both external, where it appears at the place the cervix of the womb [= the vagina] ends on the outside; and internal, where the cervix begins.[71]

Despite the care he took to be precise, the author of the *Anatomia Vivorum* found it rather difficult to draw a clear distinction between the vagina, the cervix of the uterus and the urethra. Like his predecessors, he refrained from describing the external female organs and halts 'at the place the cervix of the womb ends on the outside'.

Anatomy in the encyclopaedias

It was at this stage of anatomical science that encyclopaedias were composed with the aim of giving a literate but non-specialist audience some rudimentary information. Their great circulation, which lasted until the end of the Middle Ages and even beyond, helped to keep alive ideas which, based as they were on thirteenth-century information, were more and more out of line with the thinking of scholars.

Greatly dependent on William of Conches, the encyclopaedists of the thirteenth century parted company with him in the description of the male genital apparatus, for which they used newly available sources. The way they used these sources was not, however, always very felicitous. Thus we find, in the *De Natura Rerum* of Thomas Cantimpré, a strange claim that was to be repeated by Vincent of Beauvais: 'The man's penis is the exit for superfluous moist matter and the canal through which sperm passes; it is made of cartilage and flesh. Desire makes this cartilage appear.'[72]

This has not failed to give rise to sarcastic comments, and for malicious spirits it discredits the intellectual efforts of medieval doctors. Yet it must be pointed out that Aristotle is at the origin of this surprising claim, put forward in the *Parts of Animals*: '[The male organ] contains both sinew and *cartilage*; and so it can contract or expand and admits air into itself.'[73]

The text does indeed have *kondrôdes*, which leaves not the shadow of a doubt. Aristotle's prestige and authority ensured that this curious anatomical conception had the widest possible circulation, and no one will be surprised to encounter it again in all authors who rely, closely or not so closely, on the authority of the Stagirite.

When Albert the Great comments on this passage of Aristotle's text, his perplexity is clear: 'For at the end of the pubis is the penis in men and the opening of the vulva in women; it is certain that the penis is constituted of flesh located between the network of ligaments, arteries and veins, and of *cartilage, not true cartilage*, but rather ligaments which are hard like cartilage.'[74]

Another passage of the *De Animalibus* formulates the same reservations with regard to Aristotle's opinion.[75] This detail is of great interest for the history of the sciences. One sees how a conflict between a true scientific spirit and Aristotle's authority could be resolved in linguistic terms. All the open-mindedness of the Universal Doctor, his scrupulous attention to the information, and even his readiness to experiment, were necessary for him to dare to free himself from Aristotle's dominance.

As for Bartholomew the Englishman, he added little to the topic we are discussing. The choice of Aristotelian propositions gives far too abstract a character to the *Properties of Things*. He gets further and further away from anatomical description, and a philosophical discourse on sexual difference – in other words, the assertion of a link between form and matter – is the unavoidable conclusion.

One has to turn to literature written in the vernacular, somewhat later than this date, to find, in a work of encyclopaedic ambitions, better information on anatomy. The *Dialogue de Placides et Timéo* has the particular advantage of being very close to Avicenna's text, as close as the *De Animalibus* of Albert the Great. The author supports, for example, the doctrine of the three passages in the penis; this idea does indeed have its source in the *Canon*: 'There are three meatuses in the penis, namely the urinary meatus, the spermatic meatus, and the meatus of the *alwadi*.'[76] This last Arabic word, which includes in its root the notion of love and desire, probably designates in Avicenna something corresponding to the prostatic secretion. The author of the *Dialogue* adds a supplementary detail concerning this meatus which transports 'a thing clear as oil and white as silver'; he indicates that it also allows the 'superfluities of the humours' which are like 'a manner of filth or foam' to pass.[77] These superfluities, as the author asserts, may give rise to furuncles. In fact, he is trying to explain what modern medicine calls 'smegma'. The remark in the *Dialogue* refers us once again to Albert the Great, one of the rare authors in the Middle Ages to point out the presence of this discharge which he calls 'thickened sperm'. He refers to it, for instance, when he describes the phenomena that accompany puberty in the adolescent;[78] in it, one can recognize the spontaneous uncapping of the penis, at the time of the first erections, sometimes accompanied by a little bleeding and a more abundant secretion from the glands present in the collum glandis penis. This secretion called 'thickened sperm' by Albert the Great is indeed smegma. One can thus see that the *Dialogue de Placides et Timéo*, which, through the information conveyed and the subjects dealt with, is close to Avicenna and Albert the Great, makes available to the public not only all the traditional knowledge, but also the fruit of close observation, together with an

attempt at explanation which – even if it does not satisfy us entirely –
tackles the problems with a praiseworthy frankness.

To close our rapid review of this type of popularizing literature, we
will leap over two centuries and mention a collection of questions –
probably from the end of the fifteenth century – which, to judge from
its title, *Problemata Varia Anatomica*, seems rich in promise for our subject.
Let us take an example drawn from this text: the compiler announces
that he is going to talk about the womb. But he passes on immediately
to another question ('Why do animals copulate?')[79] followed by a
definition of coitus, its function, its consequences for the organism and
so on, without there at any moment appearing a description of the organ,
which serves merely as a pretext for these different assertions to be
brought together. The term 'anatomical' is there to introduce either a
remark on physiology, or else an account of the rules of hygiene.
Basically, it is just as if anatomy, a knowledge of which one should not
presuppose, serves merely to bring a practical discourse into line with a
tradition of scientific discussion, only the skeleton of which subsists in
the form of chapter headings. This type of scientific discourse, although
deriving its information from the greatest scholars – Albert the Great,
Constantine, Avicenna and Pietro d'Abano – constitutes merely a
collection of 'curiosities'. It is an extremely conservative form of
knowledge, since the fragments are never brought together into a whole,
and there is no synthesis which would enable one to judge of its value
as a system.

The time of anatomists and surgeons

Cautiously prefigured by autopsies for *post mortem* diagnosis, the dissection
of human corpses became a standby of anatomical investigation in
Bologna at the end of the thirteenth century. The name symbolically
attached to this new development is that of the doctor of medicine
Mondino de' Luzzi, whose *Anatomia*, completed in 1316, was to dominate
teaching for two centuries.[80] Furthermore, several years previously,
another Italian doctor, William of Saliceto, had broken new ground for
surgical literature by including, in his *Chirurgia*, completed in 1275, an
anatomical section that had hitherto been missing from this type of
work.[81] During the following quarter of a century, Henry of Mondeville
took the innovative step of illustrating his lessons with printed plates;[82]
his *Chirurgia*, composed in 1306, was translated into French as early as
1314 and was probably used in the training of non-Latinist surgeons
and barbers.

Although they were still very dependent on the sources they used, and
in particular on Avicenna's *Canon*, these authors chose their information

carefully, selecting certain things and rejecting others. Thus, they were particularly sceptical about the existence of the third passage in the penis described by Avicenna. In William of Saliceto we read:

In the penis there are at least two passages; one through which urine comes out, which is linked directly to the neck of the bladder; and another through which the sperm is expelled, which is linked directly to the sperm ducts. These two passages are joined together in the body of the penis. It is true that certain people say there is a third, through which nature expels the sperm during sleep, and which is supposed to be different from the other two, but this does not seem to me to have been verified.[83]

Henry of Mondeville later gave an account of Avicenna's theory, evincing the same scepticism; as for Mondino, he described only two meatuses.[84] The commentator on the *Canon*, Gentile da Foligno (died 1348), adduced proof by dissection in order to explain his hesitations: '*Alwadi* is the humour which is expelled when a man touches a woman; its meatus cannot be seen on dissection.'[85] For a doctor who, like Gentile, knew the *De Usu Partium*, accepting the existence of such a meatus posed a tricky problem, for it came into conflict with the opinion of the principal authority. Galen, describing the glandular bodies situated on either side of the 'neck of the bladder' and the humour similar to sperm, but more watery, which comes out of them, did not mention any duct with the special purpose of allowing this discharge to pass: 'The humour produced in those glandular bodies [the seminal vesicles] is poured out into the urinary passage in the male along with the semen.'[86] These 'glandular bodies', identified by Charles Daremberg as the prostate and, more recently, by H.J. von Schumann as Cowper's glands and by M.T. May as the seminal vesicles, were rarely given much attention by writers.[87] In the fifteenth century, Jacques Despars, endowed with a store of erudition and a perspicacious mind, enumerated 'five members serving the purpose of reproduction in the male, not including the membranes that cover them, nor the penis'. He listed the veins and arteries with their twisting passages, which correspond to the spermatic venous plexus, the testicles, the epididymis, the sperm ducts (= vasa deferentia) and glands which he described thus:

There are glands set at the root of the penis which create a certain moisture; this moisture, which is like saliva, increases the pleasure of the sexual act and prevents the duct through which the sperm comes out from being too dried up by the frequency of coitus; it also mitigates the heat and dryness of the duct through which the urine is expelled. This moisture is sometimes emitted with the sperm, sometimes without it – as when a man and a woman kiss or caress; at other times part of it gradually dissolves and part is conserved in the aforementioned glands. These glands are called, in the *De semine* and the *De Usu*

Partium translated by Nicholas of Reggio *adenes, adenosi parastate* [= *prostatas*] and *adeniformes*; for the Greek *aden* is translated as *glandula* in Latin.[88]

At the end of the development of medieval thought, after the 'introductory course' constituted by the arrival of Arabic science, Galen's text was once more intelligible, despite the obscurity of the vocabulary used in the translations. The rediscovery of the *De Usu Partium*, in a fuller and more faithful Latin version,[89] helped people to perceive the anatomical reality more clearly; this rereading, which was not yet truly critical, was made possible by the new way of seeing which the anatomists of the beginning of the fourteenth century made current. So we must look more closely at their descriptions.

The observation of human anatomy did not succeed in disproving the theory of cells in the womb; the philosophical stakes were too high for that. In his account of the dissection of two women's corpses, Mondino de' Luzzi pays particular attention to variations in the size of the uterus; he notes, with a certain astonishment, that the womb of a sow bearing thirteen piglets appeared to him, on dissection, a hundred times bigger than that of the women he had opened up.[90] He refers to the presence of seven cells, but adds the following restriction: 'These cells are merely kinds of hollow cavity existing in the womb so that the sperm may coagulate with the menstrual blood.' Likewise, Gentile da Foligno restricts the scope of Avicenna's text, according to which there are two cavities: 'There are no ventricles in the womb, but there are certain cells which do exist, even if no division can be seen on dissection.'[91] So writers continued to choose, according to the source they relied on, the number two or the number seven; or else they assumed that, in mammals, there were as many cells as there were teats; but it must be emphasized that the cells they were describing were no longer altogether the same as those referred to in the previous century.

Another noteworthy modification concerns the vein that links the womb directly to the breasts. The passage between the two was always described but, more subtly, it was seen as consisting of a network of veins and anastomoses the complexity of which had been suggested by Galen:

Now I am going to explain why the breasts have so much in common with the uteri; for this too indicates an amazing skill on the part of Nature. Since she prepared both these parts to be of service in a single work, she has joined them by means of the vessels [*aa.* and *vv. thoracicae internae*] which I have said in my discussion of the thorax go to the breasts, by bringing down veins (*vv. epigastricae superiores*] and arteries [*aa. epigastricae superiores*] to the hypochondrium and the whole hypogastrium, and then by attaching these to the vessels [*aa.* and *vv.*

epigastricae inferiores] which come up from the parts below and from which veins extend to the uterus and scrotum.[92]

Avicenna's *Canon*, used as a direct source by anatomists of the fourteenth century, repeated this theory, but, as in Galen, the account does not follow in continuous order; it starts at the beginning of the first book, in the chapter on the vena cava, and continues at the end of the third book, in the chapter on the womb:

[From a ramification of the ascending vena cava] there comes a vessel which reaches from the upper part of the chest down to the abdominal cavity . . . [Ramifications of the descending vena cava] branch out towards the muscles which are located over the stomach when the body is upright. These veins lead directly into the extremities of the aforementioned veins which reach down from the chest towards the abdominal cavity. In women, there extend from the root of these veins vessels which lead towards the womb, while the veins themselves branch out to go to the breasts; it is in this way that the breasts are linked to the womb.[93]

The idea of the existence of this communicating passage, the description of which serves only to explain the complex phenomenon of lactation, can only be sustained by going into a detailed analysis of the flow of the blood. When dealing with the links between the womb and the principal members, Mondino evinces a similar caution in referring to the link with the breasts:

[The womb] has an important link with almost all the members in the upper parts of the body: with the heart and the liver, by means of the veins and arteries; with the brain via the many nerves. In this way it is also linked to the members in the middle of the body, such as the stomach, the diaphragm, the kidneys and the peritoneum, since [these members] communicate with [the upper ones] aforementioned; it is in especial linked to the breasts as I have explained, although this link is ensured by the intermediary of other vessels which, issuing from the inferior vena cava, spring from under the clavicle.[94]

The idea that a passage exists between the breasts and the uterus, an idea originally deduced from a Hippocratic text, was thus safeguarded, but henceforth it required a sophisticated anatomical description. On the other hand, another idea, also Hippocratic in origin, namely the importance ascribed to the juvenile veins (= jugular veins) in fertility, aroused much greater scepticism. In the *Anatomia* which he dedicated in 1314 to Philip VI, King of France, Guy of Vigevano declared: 'As for me, in my time I have on several occasions cut open these veins, both in men and in women, and they have later had children.[95] Furthermore, as we shall see, the progressive abandoning of the theory which had made the brain the seat of spermatogenesis removed any usefulness these veins might have had in the process of reproduction.

Though it did not lead to any far-reaching revisions, this new way of seeing, linked as it was to the development of dissection and an extension of surgical practice, contributed nonetheless to minimizing the importance of certain archaic assertions, or to modifying the way they were interpreted. From these observations, which from now on were applied to the human being, one would be justified in expecting a correct description of the female organs other than the womb or the 'testicles'.

As far as the hymen is concerned, we will quote what Albert the Great wrote in his *De Animalibus*:

Before corruption sets in, there exist, in the cervix and at the entrance of the womb of virgins, membranes made of a tissue of veins and extremely loose ligaments which are, once seen, the proven signs of virginity, and which are destroyed by the act or even by inserting one's fingers; whereupon, the small quantity of blood in them flows out.[96]

Henry of Mondeville repeats the same kind of description, but in less detail: 'Towards the middle of the cervix one can find, in virgins, veins which are torn at the moment of deflowering.'[97] Mondino de' Luzzi claims that only one membrane acts as a sign of virginity: 'The surface [of the entrance to the womb] is in virgins covered by a subtle and venous membrane; at the moment of deflowering, it breaks, and women bleed.'[98] The word 'hymen' is not used by the anatomists of the fourteenth century; we come across it in the fifteenth century when, in his *Practica*, Michael Savonarola takes up Mondino's description: 'The cervix is covered by a subtle membrane called the *hymen*, which is broken at the time of deflowering, so that the blood flows.'[99]

The external female organs are no longer absent, as they were in the Salernitan texts, but the anatomists and surgeons of the end of the Middle Ages still show a strange lack of precision. In Mondino, we find this description of the labia:

At the end of the vulva there are two small flaps of skin which open and close over the orifice to prevent air and foreign bodies from entering the cervix or the bladder; they are like the small flaps of skin of the prepuce which protect the urethra, and that is why they are called by Avicenna the *prepuces of the womb*.[100]

This description is repeated word for word by Michael Savonarola in the fifteenth century, but the expression attributed by Mondino to Avicenna is ascribed to the author of the *Pantegni*, which is more correct.[101] In the work of the famous surgeon Guy of Chauliac, we also come across an organ similar to the prepuce, but it is probably not the labia:

[The womb] is large forsothe as the yerde [penis] turned ayen or putte withynne, in 7° De Vtilitate Particularum. It hath, forsothe, above, two cellede armes with prive stoones, as it were the purse of the prive stones. It hath a comune wombe in the myddel, as the parties of the schar bone. It hath a nekke, holow withynne, as the yerde. It hath also the prive schappe or chose [vulva] as a hellynge and mytre [glans penis and urethra]. It hath also a prive poynte [in Latin, *tentigo*] as the hole in the yerde [i.e., the prepuce].[102]

So it here seems to be the clitoris which is compared to the prepuce, as the use of the word *tentigo* – a word whose fortunes from the time of late Latin civilization onwards are not easy to follow – would seem to suggest.[103] There is a metonymic relation between this organ and the sexual ardour of women in general, which makes it difficult to interpret certain texts. With reference to the exterior orifice of the womb, Henry of Mondeville gives us the following description of the *tentigo*:

[The orifice] is made in such a way that it can open and close at any time; it is called vulva or cunt (*vulva vel cunnus*). In its middle it has a brawny membrane which hangs out somewhat and which is called by Rhazes, in the second [book] of the *al-Mansuri*, chapter seven, *tentigo*. There are two reasons for the creation of this membrane: first, it is meant to act as a passage for urine so that the latter does not spill out into the whole vulva; secondly, when the woman is sitting with her thighs parted, it affects the air entering the womb, as the uvula does for air entering the mouth.[104]

The two analogies mentioned by Mondino, Henry of Mondeville and Guy of Chauliac are Galenic in origin; in the *De Usu Partium*, the labia are compared to the prepuce, whereas a protective role similar to that of the uvula falls to the clitoris or 'nymph'.[105] On the other hand, the confusion with the end of the urethra is not Galenic. To try to explain the mistake made by the surgeon Henry of Mondeville, one can point out the inaccuracy of the sources he was using: he did not have access to the translation from Greek into Latin of the *De Usu Partium*, and furthermore, the Arabic texts are not very explicit on this topic. We have seen that 'Alī ibn al-'Abbās seems to confuse the clitoris with the labia; Rhazes is elliptic, while Avicenna and Albucasis only refer to the organ as a protuberance that has to be corrected by surgery.[106] Of course, the different names given to the clitoris did not help it to be recognized as such: Constantine the African transcribed the Arabic word as *badedera*; likewise, Gerard of Cremona sometimes used the term *tentigo*, sometimes *batharum*.[107] The explanation given by the *Synonyms* of Simon of Genoa (fourteenth century) for this Latin translation is enlightening: '*Batharum* in Arabic is a fleshy protuberance in the vulva of certain women which is sometimes so large that it is comparable to the penis.

Moschion called it *landica*.'[108] Since it appeared only in the form of a hypertrophy, the organ was doomed to clitoridectomy.

The example of the clitoris seems to us revealing – on the one hand of the place occupied by anatomy in medieval medical science, and on the other hand of the vision of woman put into circulation by that science.

It would be naive to deduce, from the absence of any description of the clitoris, or from the imprecision of such a description, that doctors were totally ignorant of woman's sensitivity. Apart from the fact that Albert the Great reports cases of masturbation, one of which very probably involves the clitoris,[109] a phrase from the *Conciliator* of Pietro d'Abano seems to us explicit: 'Likewise [women are driven to desire] especially by having the upper orifice near their pubis rubbed; in this way the indiscreet (? *curiosi*) bring them to orgasm. For the pleasure that can be obtained from this part of the body is comparable to that obtained from the tip of the penis.'[110]

This merely goes to confirm what we have had several occasions to note. Anatomical description was more often than not enslaved to a teleological interpretation, in which physiology reigned supreme, and such description was cut off not only from all observation, but from any real perception of the body – a perception other types of discourse could well show, however, even within medical discourse.

It was not until the Renaissance, and more precisely Gabriel Fallopio, that the link between a particular area of sensitivity in the female body and the presence of an organ was to be demonstrated clearly. The great Italian scientist gives us his own explanation of his predecessors' ignorance: 'This *pudendum* is so small and hidden in the fattest part of the pubis that it never attracted the attention of anatomists; it is so hidden that I was the first to discover it, several years ago.'[111] Gabriel Fallopio is here exaggerating his merits somewhat: the clitoris, as we have seen, was not completely absent from all descriptions; but it is true that, for centuries, anatomical science and lived experience never came into contact with each other.

It could also be claimed, to put it in an altogether modern way, that the medical text stuck to the limits of an exhaustively male discourse, that the model pre-existed the description of reality. The theory of the inverse symmetry of organs, as suggested by Galen, could not cope with the additional problem of the clitoris, which would upset an account in which the equivalent of the penis was the 'cervix of the uterus'; female sensitivity seemed solely vaginal. To explain the clitoris, another analogy was required, one also suggested by Galen. The opening of the woman's body constituted a danger and something to protect it was imagined;

the air which entered the body had to be tempered by an organ supposed to play the same part as that attributed to the uvula in the act of breathing. But this evident rationalization does not perhaps give a full enough idea of the image woman had in the Middle Ages; she was in league with the air and the wind. This theme, taken from the Arab world, was well received by the Middle Ages, which enjoyed repeating again and again Avicenna's example of mares being impregnated by the wind; and, in the same spirit, Albert the Great mentioned the case of a woman who derived a particular pleasure from the caress of a breath of air.[112] Certain themes represented in folklore would illustrate this female aptitude for absorbing influences and letting them circle upwards through the body. The medical practice of fumigation does not contradict this particular detail. The case of the clitoris also shows that a scientific spirit, when faced with a fact that lay outside strict logical causality, allowed to surface, or even accepted without discussion, ideas which corresponded to the deep structure of the imagination.

2

Physiology, or the Stages of Purification

Continuity and permanence were the keynotes of physiological mechanisms, as they appeared in medieval texts. The transmission of the texts of antiquity helped to constitute a set of simple explanatory principles that was scarcely questioned before the sixteenth century. The physiology of the human body, and medical treatment of it, were fitted into the traditional way the world was represented. The four elements, and the dual quality that each of them possessed, were correlated with the four humours taken from the Hippocratic school, and it was this system, developed in the greatest detail by Galen, which was to constitute the foundation of medieval science.

Although it is very well known, we will take the liberty of reproducing a diagram that is particularly suitable for showing the essential place occupied by the interaction of primary qualities (hot, cold, dry and moist) on all levels of physics and physiology (see table 2.1).[1]

One can see that the establishing of this correlation between macrocosm and microcosm brought within the doctor's reach both an explanation of the human body, and the possibility of re-establishing an equilibrium between the humours – thus restoring the temperament which an internal or external cause had broken.

THE GALENIC APPARATUS

Galen remained the intellectual master of all the doctors of the Middle Ages, even if direct transmission of his works was fragmentary until a late date. According to him there exist two sorts of *pneuma*.[2] The *pneuma zotikon* or *vital pneuma*, transmitted to all the organs of the body by the blood of the arteries starting from the left-hand side of the heart, is the active agent in breathing and combustion; it is the life-principle. The second form of pneuma is the *pneuma psychikon* (*spiritus animalis*) which

TABLE 2.1   *The interaction of the four primary qualities*

North

COLDNESS	Water Winter Phlegm Phlegmatic		MOISTURE
West	Earth Autumn Black bile Melancholic	Element Season Humour Temperament	Air Spring Blood Sanguine	East
DRYNESS	Fire Summer Yellow bile Choleric		HEAT

South

fills the brain and its lobes but which is not, however, the soul. It is the product of the flow of blood to the brain. The brain and the entire nervous system are nourished by the veins and the arteries; the veins transport nourishment, the arteries the vital spirit (*pneuma zotikon*). Galen also supposes that air can reach the brain through the nasal cavities, which gives this organ a relative independence from the action of the heart, arteries and lungs.

The vital *pneuma* is drawn from the air inhaled during breathing; the psychic *pneuma* is formed from the vital *pneuma* in the choroid plexus located at the base of the brain, when arterial blood passes through the brain's vessels. This process of distillation by circulation in a network of capillaries is altogether analogous to the process that produces dealbation of the blood in the genital organs. The question of the transmission of *pneuma* through the body is of primary importance: authentic Galenic doctrine makes of the nerves a compact assembly of fibres, whereas in Galen's continuators – and especially in the writers of the Middle Ages – they are hollow ducts. Another difference between medieval Galenism and the original work of the Greek doctor resides in the recognition of a third *pneuma*, called 'natural' (*spiritus naturalis*), which has its seat in the liver and ensures vegetative functions. This modification of the doctrine came about, it seems, as early as antiquity;[3] several Arabic texts, despite being profoundly marked by Galenic influence, imposed on the Western Middle Ages this tripartite division between spirits and the functions that they perform.[4] It was all the easier to adopt this notion since it corresponded to the tripartite division of the soul in Plato and Aristotle.

In this theory, the sperm is a fluid that is not an integral part of the system of the four humours; it is a specific product which collects in both male and female testicles. As we have already said, the latter clearly represent the ovaries, which Vesalius still called *testes muliebres*, before Fallopio rejected the function that was ascribed to them under this name. From Galen the Middle Ages took the doctrine of the two sperms. They are produced from the blood of the veins and the arteries, when they reach the glands:

In this interweaving the blood and pneuma passing to the testes are very greatly concocted, and it is possible to see clearly that the humor contained in the first coils is still like blood and that in the succeeding coils it keeps getting whiter and whiter until in the very last ones, that end in the testes, it has been made absolutely white.[5]

Galen also thought that in women, the ovaries sent the sperm into the uterus through the ducts (the Fallopian tubes) which played a similar part to that supposedly played by the epididymis in the male organs.

Galen took from his predecessors the theory of the opposition between left and right, a determining factor for the sex of the future embryo, which comes from one or other side of the father and is nourished in one or other part of the womb. His description of the anatomical structure of the genital organs, organized around this opposition, shows an accuracy that was never surpassed or even equalled in the Middle Ages. Everything is constructed around the liver, that is, the part of the body from which the veins spring:

Well, then, where the vena cava first arises from the liver and, still suspended, bends down along the spine, it has the right kidney lying to the right of it and next, a little lower down, the left kidney lies on its left. . . . Of the vessels that pass to the generative parts, however, the ones [*a.* and *v. ovarica, a.* and *v. spermatica interna*] going to the right uterus and right testis start from the great vessels themselves that are along the spine, the vein from the vena cava and the artery from the great artery, but those that reach the left testis in the male or the uterus on that side in the female (and there are two of these vessels, one artery and one vein) do not start from the great vessels themselves, but from the vessels passing to the kidneys.

Hence it is clear that the left testis in the male and the left uterus in the female receive blood still uncleansed, full of residues, watery and serous, and so it happens that the temperaments of the instruments themselves that receive [the blood] become different. For just as pure blood is warmer than blood full of residues, so too the instruments on the right side, nourished with pure blood, become warmer than those on the left.[6]

The reader must excuse us for this rather long quotation, but it enables us to show that the Middle Ages, in so far as they inherited Galen's ideas, did not only accept, with the opposition between left and right, a dividing up of space that was associated with various taboos, but also a scientific explanation of the human body with arguments and demonstrations that were to be questioned only by a thoroughgoing revision of the physiological system. The explanation that we have quoted does indeed rest on an accurate piece of anatomical observation: the right spermatic vein (issuing from the anterior plexus) and the right ovarian vein (issuing from the pampiniform plexus) end in the inferior vena cava, whereas their counterparts end on the left in the renal vein.[7] On the other hand, the consequences drawn from this observation were merely a result of the strict application of the principles that control physiological processes: the purification of the blood by a succession of coctions, and the interaction of the four primary qualities and the characteristics attached to them.

Both logic and the authority of Hippocrates again reinforced the prerogative that Galen granted to the right-hand side: 'And further, these parts had the advantage in position from the very beginning; for I have demonstrated many times the correctness of Hippocrates' statement that parts lying in a straight line necessarily get greater benefit from one another.'[8]

Furthermore, since they were linked directly to the liver, which was naturally hot, the right womb and the right testicle were much hotter than their equivalents on the left, and were thus suited to produce and nourish the male. The data of Hippocratic authority were respected and observation confirmed traditional learning. The assertion that woman was necessarily of a colder nature than man was to be the obligatory conclusion in medieval disputes about sexual differentiation;[9] the overlapping of anatomical, physiological and philosophical explanation on the one hand, and the traditional high value placed on the right-hand side and the quality of heat on the other, left little room for arguments to the contrary.

Galen's contribution to knowledge of the circulation of the blood between the mother's organism and that of the embryo, as well as to the problem raised by the breathing of the foetus, was on a level of scientific and medical thinking which medicine reached but rarely in the Middle Ages. These theories belong more to the field of gynaecology and are of only tangential interest to our enquiry into sexuality. It will have been noticed that the theory concerning the modification of the blood's circulation, that is, the transformation of menstrual blood into milk for the suckling baby, constitutes the only discussion of any importance devoted to the breasts. There is indeed, between the breasts and the womb, a close sympathy which anatomical description has already enabled us to mention. According to Galen, the menstrual stream is evacuated every month by means of the vessels that lead to the womb, and these same veins serve the purpose of bringing nourishment to the foetus during pregnancy. When the child has been born, all the menstrual blood, by a modification in the circulatory system that intrigued medieval authors, flows back to the breasts. The affinity between milk and menstrual blood, asserted by Hippocrates, and repeated in greater detail by Galen, was an idea which the Middle Ages referred to again and again. As we have already noted, this notion was present in the *Etymologiae* of Isidore of Seville: 'The blood used for the nourishment of the uterus goes to the breasts and takes on the quality of milk.'[10] Theories so constantly reworked were powerful factors in the way the world was represented, factors that were never questioned for almost five centuries. From the simple explanatory schemes that lay behind these medical theories, such as purification and elaboration through circulation in the canals, were born many other strange ideas and dreams of the imagination, always within the general framework of a permanent interplay between microcosm and macrocosm. Whereas the blood that had reached a definite stage in its purification became semen, milk was formed from the menstrual blood. A celebrated distich from the School of Salerno stated, furthermore, that the first stage of the embryo was 'like milk' before the blood appeared:

> Conceptum semen sex primis crede diebus
> Est quasi lac, reliquisque novem fit sanguis.[11]

All the mystery of the appearance of life turned on the interaction of these two fluids of two different colours that were each meant to serve different purposes. This raises for us the question of the origin of the liquids of life, the origin of semen.

THE PRECIOUS LIQUOR

The origin of semen was one of the questions to which the Middle Ages replied in the most ambiguous way possible. Even after the concept of haematogenesis – one of the main ideas spread by Arab doctors – became generally accepted from the twelfth century onwards, resurgences of the encephalomyelic idea could still be found, as well as the different theories developed in antiquity, all of which coexisted. Thanks to the work of Erna Lesky, there are few obscurities left as regards these theories, and the three Greek doctrines, as well as their principal representatives, are well known.[12] The connection between brain, spinal cord and semen (and its affinities with the cosmos) seems to have originated in the beliefs of ancient Persia; this hypothesis reveals both the close relationship that exists between procreation and religious thought, and also the fact that beliefs concerning sexuality and gynaecology belong to the most remote substratum of ethnic groups. The earliest of Greek theories located the origin of sperm in the brain, an idea supported by Alcmaeon of Crotona as well as by the Pythagoreans, and among them, Diogenes Laertius, for whom semen was a drop of the brain (*stagōn enkephalou*). A proof that was long used to support this thesis could be found in the Hippocratic writings, where reference was made to 'Scythian eunuchism' or 'women's disease', a disease which one could survive so long as the veins located behind the ears were cut.[13] This led to a flow of blood and the patient's losing consciousness; some people came round in good health, others died. In any case, sterility set in after the operation. This story helped to shore up the theory which, locating the origin of sperm in the brain and spinal cord, had it travelling to the testicles through two veins lying along the spinal column. Plato in the *Timaeus* voiced an opinion close to this when he declared;

[The gods] bored a hole into the condensed marrow which comes from the head down by the neck and along the spine – which marrow, in our previous account, we termed 'seed'. And the marrow, inasmuch as it is animate and has been granted an outlet, has endowed the part where its outlet lies with a love for generating by implanting therein a lively desire for emission.[14]

The second theory, associated with the names of Anaxagoras and Democritus, had the sperm coming into existence in all parts of the body; this theory also figured in the Hippocratic Collection, which in this case presented the medieval supporters of pangenesis with an additional argument, by asserting that since sperm comes from all parts of the body, it is 'weak when it comes from the weak parts and strong when it comes from the strong parts'.[15] This was used to explain the way certain infirmities were inherited. This theory could, furthermore, be combined with that of encephalogenesis; the humour was formed in the brain, then travelled to the genitals and throughout the whole body, but especially into the spinal cord, which was the central storehouse and the organ of distribution.

While Isidore of Seville adopted in broad outline this latter variant,[16] William of Conches proposed in the *Dragmaticon Philosophiae* a solution more clearly tending towards pangenesis: 'Sperm is thus the seed of the man, composed of the purest substance of all parts of the body.'[17] The form of the dialogue leads the master to put forward the argument that certain ailments can be inherited, such as chiragra or podagra. The disciple immediately raises the objection: 'If someone has his hands, feet or ears cut off, his child will nevertheless be well-formed.' This argument is incapable of shaking the master's certainty, for, according to him, nature avoids imperfection and develops the part of the body that should logically be missing in the embryo by using a similar part of the body. This sketch of a debate shows that the importance of an apparently innocuous question, such as the origin of sperm, was clearly perceived: in fact the question of heredity as well as the immediate links between parents and children – in other words, doubtless one of the major components in the emotional life of families – were at stake.

The most serious challenge to ancient pangenetic theories was encountered by the Middle Ages in a form invested with all the authority of Aristotle. The treatise *Generation of Animals* insists:

Our own statement must therefore be the opposite of what the early people said. They said the semen is that which was drawn from the whole of the body; we are going to say the semen is that whose nature it is to be distributed to the whole of the body. And whereas they said it was a colliquescence, we see it is more correct to call it a residue.[18]

For Aristotle, sperm was a residual product and the excretion of sperm took place via the uterus, the sexual organs and the breasts. The formula remembered and developed in the Middle Ages was thus the following: sperm was 'a residue derived . . . from useful nourishment in its final form'. As blood was itself a product of food, sperm derived from food could be nothing other than blood, a substance analogous to blood, or a product that came from blood. That is why it was altogether possible to predict that sexual excess would lead to a discharge of blood.

Reading works inspired by Galen, scholars of the Middle Ages discovered another topic for further thought – namely, the part played by the testicles (which, let us recall, were counted among the 'principal members') in the formation of sperm. Contrary to Hippocrates and Aristotle, who saw them as mere receptacles, Galen ascribed a specific function to them. The raw material of sperm was, to be sure, provided by blood at a definite stage in its coction, but this blood underwent a transformation, shown above all by a progressive dealbation, in the circumvolutions formed by the spermatic veins and arteries (pampiniform plexus), and it turned into semen on arrival in the testicles.[19]

Arab medicine shed hardly any light on the confusion that the coexistence of different ancient theories created. Under Avicenna's authority one can read in the first book of the *Canon* that 'The sperm has its principle in the humours', and, in the third book:

The sublime God created the testicles to be, as you know, the principal members which engender sperm from the moisture that is brought to them in the veins; this moisture is like the residue of the food that has reached the fourth stage in the whole body. It is a better-digested and subtler blood.[20]

One of the commentators on the *Canon* did not hesitate to point out the contradictions in this account: how could a residue from the fourth coction be subtler than the blood which came from the second coction?[21] If the question of the origin of sperm was debated by the doctors of the Middle Ages with less passion than that of the existence of female seed, this is perhaps because no theory succeeded in compelling recognition as a coherent whole and, because of this, compromises were possible. Thus the *Conciliator* of Pietro d'Abano, a basic work from the beginning of the fourteenth century onwards, concluded that the sperm descended directly from the testicles and the seminal vesicles, but indirectly from the whole of the body.[22] Among the arguments in favour of pangenesis, one finds the remark that the pleasure which comes from emission spreads throughout all the parts of the body.

The acceptance of haematogenesis did not succeed, either, in completely effacing every reminiscence of the theory that located the origin of the sperm in the brain: too many beliefs relative to sexual excess were involved in it, beliefs that were circulated by the most eminent scholars. Thus Albert the Great, in one of the *Questions on Animals*, attempted to decide whether sperm, a residue of the final digestion, came more from one part of the body than from another.[23] After examining the arguments *pro* and *contra*, he maintained that sperm came more from the principal members, and especially from the brain, whose substance, in its whiteness, softness and moistness, corresponded to that of semen. In support of this conclusion, he related an experiment and an anecdote. The former demonstrated the close relation linking the brain to the testicles; bathing the genitals of a drunken man brought him round from his drunken stupor. The anecdote that followed this 'experiment' may be considered as an *exemplum* designed to arouse fear in the debauched. Albert the Great specified that it was reported to him by master Clement of Bohemia: so we are not dealing with a first-hand eyewitness account, but the story struck Albert as sufficiently veracious to be mentioned, all the more since it most likely concerned a monk from a rival order. This man, described as 'half-starved', died after having 'desired' a beautiful lady seventy times before matins was rung. The autopsy, carried out because of the nobility of the family the monk came from, revealed the emptiness of his brain, which had been reduced to the size of a pomegranate, and the fact that the eyes had been destroyed. Albert the Great concluded: 'This is the sign that coitus drains, above all, the brain.'

The vulnerability of the eyes, organs closely linked to the brain, formed another topos of medieval medical literature to which the authority of Aristotle gave all its force. This theme was expressed not only in the *Problems*,[24] but also in the authentic treatise *Generation of Animals*. For Aristotle, the residue which formed the sperm was not supplied uniformly from all parts of the body:

For of all the regions in the head the eyes are the most seminal, as is proved by the fact that this is the only region which unmistakeably changes its appearance during sexual intercourse, and those who overfrequently indulge in it have noticeably sunken eyes. The reason is that the nature of the semen is similar to that of the brain; its matter is watery whereas its heat is a mere supplementary acquisition.[25]

Aristotle seems thus to have lent his authority to an argument that was particularly effective in repressing sexuality. One cannot fail but think of the alleged blindness brought on by masturbation, a spectre that certain nineteenth-century doctors summoned up a little too frequently. In the Middle Ages, the deterioration of the eyesight was constantly being mentioned as part of the damage caused by coitus – both in scientific treatises and in works of a more popularizing kind; thus the *Problemata Varia Anatomica*, written in the fifteenth century, claims: 'Coitus destroys the eyesight and dries up the body.'[26]

Restricting ourselves as we are to the field of sexuality, we cannot tackle the important medical and philosophical debates that stood out as milestones in the history of theories of reproduction. So we will merely point out the different types of questions that were thrown up when people reflected on the nature of sperm, and the effects this thinking had on the role ascribed to each of the two sexes. It is interesting to survey the state of knowledge in the person who often, in his doxographies, brought together the most information. Citing Aristotle as his authority, Vincent of Beauvais repeated the different ideas that we have already been drawing on. But he added the following: 'That is why the warmth of the sun and animal warmth are present in sperm.'[27]

This raised the question of the influence of the heavenly bodies on the reproductive process of the human species and the philosophical debate relating to the *rationes seminales* that we will attempt to summarize.[28] Saint Augustine explained the evolution of certain organisms by the action of powers present in the world from its creation onwards. This tradition, followed by thinkers such as Saint Bonaventure, Albert the Great and Roger of Marston, thus implied that a potential power lay buried in the *materia prima*. But Aristotelian thought, as it was transmitted by the Arabs and interpreted by Saint Thomas Aquinas, tended to reaffirm the exclusively passive nature of the *materia prima* – and thus contradicted the theories of *rationes seminales*. These questions were among the propositions condemned by Etienne Tempier in 1270 and 1277, during the crises that were shaking the university.[29] Without going into all the complexities of the debate, we can say that compromise solutions could exist between supporters of the inert-matter theory and those who supported the decisive role played by an external action. One of these solutions, quite an ingenious one, was put forward by Robert Kilwardby: the vegetative and sentient soul relied on operations intrinsic to matter, whereas the rational soul had to do with an external influence.[30] Thus, for instance, Thomas Aquinas included in an enlarged conception of the *rationes seminales* the active virtues of the heavenly bodies.

For Saint Thomas, the seed is a recipient of the power of the heavenly bodies, through which God exercises his action on the world.[31] Natural heat (*calor elementaris*) is reinforced by the heat of the sun, so that the sperm contains a threefold heat: the elemental heat of the semen, the heat of the father's soul and the heat of the sun. The sun's action can be described more precisely in the following way:

For while the father is the reason for the generation of a particular newborn child, he is not the reason for its generation as a member of the human species. The univocal causality exercised by the father has to subordinate itself to a universal causality, that of the heavenly bodies and notably of the sun.[32]

The success of the act of generation depends, according to Aquinas, on the action of three agents: the father who accounts for individuality, the heavenly body which accounts for the fact that the child belongs to the human species and lastly the angel who accounts for its corporeal nature, for the capacity matter has to receive the rational soul – the angel, that is, who enables matter and spirit to be yoked together. One can also see that in these few statements is raised the question of the legitimacy of astrology.

The mere allusion to the sun that one finds in the doxographies of Vincent of Beauvais enables one to account for the importance of philosophical problems raised by the question on the nature of sperm. Sexuality and embryology thereby came to occupy the very heart of medieval thought, and on this topic the best authors deployed a subtlety and depth of analysis that have rarely been equalled.

altam pfundicatem expositionis libror.
ut pdixi sentiens. urribusq; receptis de
egritudine me erigens uix opus istud
decem annis consummans ad finem
pduxi. In diebus autem HEIYRICI
yoguntum archi epi ⁊ Conradi roma
nouum regis ⁊ Cunonis abbatis in
monte beati DYSIBODI pontificis
sub papa Eugenio he utsiones ⁊ uerba
facta sunt. Et dixi ⁊ scpsi hec ñ secundu
admuentione coedis mei aut ullius ho
minis. sed ut ea in celestib; uidi. audiui
⁊ pcepi p secreta misteria di Et iterum
audiui uocem de celo michi dicentem.
Clama ⁊ scribe sic.

Incipiunt capitula libri SCIVIAS
SIMPLICIS HOMINIS.
Capitula pime utsionis pime partis.
I. De fortitudine ⁊ stabilitate eciuitati
regni dei.
II. De timore domini.
III. De his qui paupes spu sunt.
IIII. Quod uirtutes a do uementes. tmites sm
⁊ paupes spu custodiunt.
V. Quod agnitiom di abscondi ñ possint
studia actuum hominum.
VI. Salemon de eadem re.

FIGURE 2.1 *A scene showing the soul being infused into the body. From Cod. B, Ms of Scivias d'Hildegarde de Bingen, Hessische Landesbibliothek, Wiesbaden. (Photo. Rheinisches Bildarchtv)*

Aristotelian thought gave a very philosophical twist to thinking about sexuality and embryology. The finest example of this way of thinking is the *De Formatione Corporis Humani in Utero* written in 1276 by Giles of Rome.[33] For him, the male sperm imparts movement and force (*virtus*) during the formation of the embryo. It should be compared to a carpenter, and the menstrual blood represents the wood on which its activity is exercised. Giles wonders what the real nature of sperm may be and is astonished at its power, so great is the diversity of bones, nerves and other parts of the body. Aristotle's authority enables him to identify the virtue present in semen with the divine virtue, then to establish a likeness with the divine intellect, so as to be able to assert that sperm has in it something of a separate substance, whereby it is placed far above matter.[34] In this notion we again come across the question of the intermediaries between the will of God and bodily substances; all this forms an elegant solution designed to account for the way the human being is 'programmed'. The danger of Aristotelianism and its avatars is that it introduces a complete imbalance between the role played by woman and that played by man.

If we had to recapitulate all the qualities ascribed to the male sperm by Giles of Rome at the end of the thirteenth century, we would again talk of its heat; produced by the man, in whom heat is the dominant factor, heat is sperm's essential quality. The woman, who does not possess this quality to the necessary degree, supplies only imperfect products, which she will not be able to bring to the final stage of their elaboration, to the last degree of coction. One has also to remember that sperm has the property of coagulating the liquids that will give rise to the first state of the embryo. It behaves like rennet on milk – an action analogous to that of an enzyme, as Mr Anthony Hewson points out.[35] This analogy also exists in Indian embryology, it is evidently to be found in Aristotle and it is mentioned in the Bible; it represents a kind of fundamental analogy, which belongs to the treasury of pre-scientific thought, and so we will not be very surprised to see it cropping up again to characterize the operations of the alchemists.

In summary, the most important elements in the representation of sexuality and gynaecology are the *pneuma* and the heat present in the male semen. For the Pythagoreans, the man's sperm contains a hot breath (*thermos atmos*). Heat is an attribute that can easily find its place in the interplay of qualities that contrast female and male; this same quality plays a determining role in the sexual differentiation of the embryo, throughout antiquity and the Middle Ages. Aristotle draws up a very full inventory of the properties of sperm, which, partaking of all the elements, also contains hot air; the description concludes with a metaphor: 'That the natural substance of semen is foam-like was, so it seems, not unknown even in early days; at any rate, the goddess who is supreme in matters of sexual intercourse was called after foam.'[36]

While all the theories applied to spermatogenesis are represented in the Middle Ages, the pneumatic component is, whatever the theory adopted, clearly in evidence. As an example, here is what Nemesius of Emesa said, translated for the first time in 1085 by Alfanus, and then by Burgundio of Pisa in 1193:

Indeed, the organs of reproduction are, in the first place, the veins and the arteries; it is there that semen is produced from blood, just as milk is produced in the breasts. This semen, passing through various roundabout ways, is sent first of all to the head; from there, it descends towards the sexual organs through two veins and two arteries. By cutting the veins located to the side of the ears or those which are near the carotid arteries, one makes a man incapable of begetting children. These veins and these arteries become a tortuous and varicose network next to the scrotum, where the moist sperm is poured into each of the testicles. It is here that the sperm undergoes one final metamorphosis: the moist sperm accompanied by pneuma is ejaculated by the varicose epididymis that lies behind the testicles. This presence of pneuma in the sperm is explained by the fact that it is an artery that emits the semen.[37]

The opinion of Nemesius is of great value: it raises in exemplary fashion the question of the circulation of the blood. For him, semen comes from veins that transport the blood (this is evidently so, since sexual excess leads to pure blood being ejaculated) and also from arteries that contain only *pneuma*. In the ideas of Nemesius, the heritage of the Hippocratic tradition and the weight of Aristotelian and Galenic theories are mixed together. They thereby illustrate the complexity of the legacy of antiquity, a complexity that medieval authors seemed to cope with quite easily. Works of popularization did not bother themselves with a theoretical substratum like this and asserted peremptorily with William of Conches: 'There are three things necessary [for reproduction]: the seed which is emitted, the heat which inflames the man and serves to extract the seed (for we know that cold freezes the humour), and the pneuma [*spiritus*] which stretches the penis and expels the seed.'[38]

It seems that the importance of these discussions on the nature of sperm was purely intellectual. All agreed to see in it 'man's purest blood', the residue of the last coction; that is, all the theories were so slanted as to prove the pre-eminence of the product formed by the male. But female seed was at the centre of a much more bitter quarrel, for the role ascribed to woman in reproduction was part of the setting-up of a hierarchy which, mediated by scientific and theological thought, was not without its effects on society.

THE INDEFINABLE ADDITIVE

The polemic concerning female sperm lasted from beginning to end of the Middle Ages. This idea, which has succeeded in arousing the mockery of certain historians of the sciences, was part of the tradition of Indo–European thought. Vedic literature, complemented by Brahmanic texts, mentioned female sperm. This is the way the mother's contribution to impregnation is referred to: 'The wife clasps the husband in her arms, they spread the virile milk, in delivering herself she milks for herself the rása.' According to Jean Filliozat, there was a correspondence between the mechanism of the reproduction of living beings and the structures of the cosmos. This passage from the Vedic texts must, he says, be compared with those in which a seed (*retas*) is ascribed to the earth-mother as well as to the sky-father.[39]

This conception of the role played by woman returns in ancient Greece. Censorinus brings up the question in his work on gynaecology (*On the Day of Birth*), ascribing the theory to Alcmaeon, Democritus, Anaxagoras, Empedocles and Parmenides.[40] The last-named, in a didactic poem on nature, points out that the embryo comes into being through the mixing of the seeds: their opposition is a way of explaining how the foetus's sex is determined.

We cannot here retrace, in all the Greek doctors and philosophers, the history of an idea that was often developed with much subtlety. To summarize, we will say that the supporters of female sperm came up against the authority of Aristotle, who denied most categorically the existence of a female seed. The treatise *Generation of Animals* includes a definitive assertion:

Now it is impossible that any creature should produce two seminal secretions at once, and as the secretion in females which answers to semen in males is the menstrual fluid, it obviously follows that the female does not contribute any semen to generation; for if there were semen, there would be no menstrual fluid; but as menstrual fluid is in fact formed, therefore there is no semen.[41]

Aristotle even replies to the objections of the supporters of female sperm: 'There are some who think that the female contributes semen during coition because women sometimes derive pleasure from it comparable to that of the male and also produce a fluid secretion. This fluid, however, is not seminal; it is peculiar to the part from which it comes in each several individual.'[42]

This secretion is specific to those women who are 'fair-skinned' and does not take place 'in dark women of a masculine appearance'. The categories of physiognomy enable one to explain both the presence of an emission whose nature is not completely specified, and also its absence. Aristotle's procedure consists in treating the phenomenon as if it were a variable, dependent as much on the individual woman as on her diet. In the Middle Ages the full force of this doctrine evidently coincided with the appearance of translations of Aristotle's works, in the middle of the thirteenth century.

An important element in the resistance to this opinion is Hippocrates' clear and unambiguous position on the subject: 'The woman also ejaculates from all her body, sometimes in the womb – and the womb becomes moist – sometimes outside if the womb is open wider than should be the case.'[43]

There is for him no doubt about it; the embryo is indeed constituted by the union of two seeds. Galen, so often adopted and commented on by the Arabs, must also be ranged among those who defend the idea of female sperm, although his opinion is more discriminating: 'Besides contributing to the generation of the animal, the female semen is also useful in the following ways: It provides no small usefulness in inciting the female to the sexual act and in opening wide the neck of the uteri during coitus.'[44]

He goes on to specify that it is from the sperm that the allontoic membrane is formed. In his systematic study of the secretions, Galen discusses the liquid 'produced in those glandular bodies' in male and female, which in the latter flows out through the vagina. Nonetheless, for him, male superiority is well established, since the female sperm in no way approaches in quality that of the male; at most it can be compared to the prostatic liquid.

Nemesius of Emesa, who was known to the Middle Ages as early as the eleventh century, had already emphasized the opposition between Aristotle and Galen. Aristotle and Democritus, he said, did not want to grant any role in reproduction to the female sperm. But Galen, criticizing Aristotle, declared that female emission and the intermixing of the two sperms were necessary. Despite being acquainted with these two antagonistic positions, Western scholars were until the thirteenth century faithful to the notion of female sperm which they had taken from Arab medicine. 'Ali ibn al-'Abbās reworked the theories of Hippocrates and Galen, and presented the female seed as a sort of diluent, whose action was indispensable:

The mixing of the two sperms is necessary for two useful purposes. The first is that the woman's sperm is a suitable source of sustenance for the sperm of the man, because the sperm of the man is thick and of a hot constitution, whereas the sperm of the woman is thin and of a cold constitution. Because of its thickness, the sperm of the man cannot spread out sufficiently and through its heat it would spoil the substance of which the foetus is made; the sperm of the woman is thus necessary to moderate its thickness and heat. Its second use is the formation of the second membrane surrounding the foetus. For the man's sperm, moving forwards in a straight line, does not reach the horn-like extensions and does not spread out over the whole internal surface of the womb. So the sperm of the woman is necessary to reach the places where the sperm of the man has not reached.[45]

The text taken from Galen continues with the celebrated analogy between the baking of the cake called *itrion*, which can easily be detached from the bronze receptacle, and the membrane which surrounds the foetus. Thanks to the female sperm, the difficulty represented by the appearance of a container within another container is resolved. Avicenna gives the same details, but is more specific:

As soon as the two seeds have been mixed, the ebullition of which we have spoken takes place and the swollen part and the first membrane are created; then all the sperm is suspended from the horn-shaped protuberances, and it there finds its nourishment as long as it is still sperm, until it starts to draw its nourishment from the menstrual blood and from the cavities [the openings of the veins] to which the membrane which has been formed is attached. According to Galen, this membrane is like a protective coating left behind by the sperm of the female when it flows towards the place where the male's sperm also flows, and if it does not join with the sperm of the male at the very moment it is shed, it nonetheless mixes with it during intercourse.[46]

All this can clearly be read in the medieval encyclopaedists and in particular in Bartholomew the Englishman in John Trevisa's translation into Middle English, which we cite here:

Aboute the getynge and generacioun of a childe, it nedith to have covenable mater, and spedeful place, and service and worchinge of kynde, [hete], as cause *efficiens* 'and worchinge and doynge', and spirit that geveth vertu to the body and governeth and reuleth that vertue. The mater of the childe is mater *seminalis*, that is ischad by worching of generacioun, and cometh of alle the parties of the fadir and the modir. First this mater is isched in the place of conceyvynge abrood, that is by the drawinge of vertue of kynde igadred togedres in celles of the modir, and is medled togedre by worchinge of kinde hete. For but digest blood of the fadir and of the modir were imedled togedre there mygte be no creacioun nothir schapinge of childe, for the mater of blood that cometh of the male is hote and thicke, and therfore for the grete thicnes it may nougt sprede itself abrood. And also for passinge hete the mater of the childe schulde be distroyed and wastid but it fongith temperament of womannes blood, that hath contrarie qualite[es].[47]

In order to convince his readers of the existence of female sperm, William of Conches puts forward as an example the case of a raped woman. This argument, repeated several times, in particular by Vincent of Beauvais, makes it appearance in the course of a long argument, of great importance for the history of sexuality in the Middle Ages. A precondition of the woman's being fertile is pleasure, which leads to the emission of seed and thus to impregnation: 'For prostitutes who have sexual relations for money alone, and who take no pleasure during the act, have no emission and thus do not conceive.'[48]

The prostitute who loves only one man can again enjoy pleasure and can once more become fertile – though this is rare, adds the author. It would be difficult to give a better explanation of sterility, doubtless a punishment for repeated infections. Let us turn to cases of rape: the women involved have had no pleasure during the act, and have not emitted any seed – and yet it sometimes happens that they conceive. To this objection, William of Conches replies with a theologically inspired pessimism that could be seen as a terrible masculine cynicism: 'Although in rape the act is distressing to begin with, at the end, given the weakness of the flesh, it is not without its pleasures.'[49]

All the ideas of William of Conches were in wide circulation: we have noticed, in the *Speculum Naturale* of Vincent of Beauvais, that almost all the sections of the *Dragmaticon* relating to gynaecology and sexuality are cited; they also feature in a pseudo-treatise on gynaecology, ascribed to Albert of Trebizond, which is merely a compilation without any guiding idea;[50] we meet with them again in a manuscript of the *Prose Salernitan Questions*, edited by Brian Lawn. On the topic of female sperm, we learn from this latter text that, in the case of adultery, when two sperms come together, the influence of the male is predominant, since the breath (*spiritus*) of the father proceeds from a greater strength of will and greater mental powers.[51] In other parts of this same text, one finds more commonplace ways of applying the doctrine, since what is in question is the determining of the sex of the embryo in accordance with the quantity of sperm and the place the seed settles in the womb.

In the medieval encyclopaedias, the defence of female sperm could be presented with a certain aggression. Such is the case with Thomas of Cantimpré: 'And thus certain people say that only the virile seed is necessary for conception and that the female seed is not. Those who put forward this opinion are merely telling lies.'[52]

Other authors avoid polemics and merely set down the opinions of the authorities side by side. It is clear, from the texts we have examined, that the emission of female sperm is closely associated with the mechanism of pleasure in the woman. It is a question that we will have to return to, when we analyse the mechanisms that come into play in the sexual act.

In the opposite camp are found the defenders of the Aristotelian doctrine. For them, the existence of female sperm is not denied, but its role in the formation of the embryo is reduced. It is on this subject that Giles of Rome, at the end of the thirteenth century, composed a meticulous account in the treatise already mentioned, *De Formatione Corporis Humani in Utero*. The text is difficult to read; it keeps coming back to the same subject, but each time it says noticeably different things.[53] The work's purpose is to reconcile the doctors' point of view with the arguments of the philosophers; to refute Galen and Avicenna, it invokes the aid of Averroes, who helps to back up Aristotle's positions. The opinion that principally needs to be refuted is that of Galen, who claims that female sperm in combination with male sperm has a certain amount of formative virtue. Giles of Rome follows Aristotle in his erroneous interpretation of the vaginal discharge. If one accepts the anatomical description of Aristotle and Galen, that is, the existence of the ovaries, which are imagined to be similar to the male testicles, the principle of teleology is to a great extent found wanting. The formative virtue belongs to the male sperm and the menses are the matter on which it operates. The resemblance between parents and children comes not from the proportions in the mixture of the two seeds, but, according to Aristotelian doctrine, from a greater or lesser resistance on the part of the matter opposing the action of the formative virtue. All the efforts of Giles tend to leave only one agent responsible for creating form, and do not admit of any hierarchy between the two seeds, for such a presentation brings the female sperm onto the side of the form-giving agent. From the moment one supposes that female sperm possesses the power to bring order into matter – as Galen had done – there are in women two humours capable of taking part in reproduction: the one actively (the sperm) and other passively (the menses). Thus, if one reasons in philosophical terms, women ought to be capable of conceiving all by themselves.

Discovering the function of the female sperm is a difficult business: it could be matter, a second matter, going into action at the same time as the menses, when the embryo is formed. The female sperm would thus be responsible for the formation of the bones, nerves and arteries, while the menses would be suitable for forming the flesh and the fat. This is an opinion that has to be refuted: the female sperm is no more endowed with active virtue than it is matter which is necessary for creating the foetus. It is a humour half-way between sperm and menses, similar to the male secretions (other than the emission of sperm) – that is, analogous to the prostatic liquid, or even a liquid 'half-way between water and sperm'.

In the reasoning employed to disparage systematically the female seed, a number of arguments play a part, in particular the claim that the woman should secrete more sperm when she is pregnant so as to provide the foetus with food; since this is said not to be the case, this substance clearly plays no part as matter. Nor does the female sperm exert any action on the menses, for it would then possess an active virtue incompatible with its nature as matter. The conclusion of the whole dense and involved demonstration is that female sperm plays no role, not even a passive role, as a complement of the menses, and that it does not play any part in inducing form. It is thus not necessary to conception.

Even more statements are brought in to back up this conclusion. They provide a wealth of information on beliefs relating to sexuality in the Middle Ages. A woman can become pregnant after an incomplete sexual act: the text mentions *coitus interruptus*, that is, a male emission 'outside the vessel', and clearly without the woman having had an orgasm.[54] Averroes quotes the example of several women who had become pregnant without emission, and adds a decisive argument, relating the case of one of his female neighbours who was impregnated by the bath-water in which a man had shed his semen.[55] Thus, the woman's pleasure is in no way required since the vulva possesses the specific property of attracting the sperm even without the sexual act being performed. The story of this 'neighbour' of a great scholar was repeated by all authors up to the end of the Middle Ages: the more Aristotelian believed it, while the supporters of the Galenist solution made the most of this opportunity to poke fun at the credulity of philosophers.

On the anatomical level, the function of the ovaries, which Aristotle did not know existed, raises a problem, for nature creates nothing in vain. Averroes merely establishes a parallel with the male mammary glands. Giles of Rome is unclear on this point: in his treatise on embryology, he explains the sterility of women in whom the ovaries are missing by the fact that they are deprived of the heat produced in these organs; in the *Commentary on the Sentences*, he follows Aristotle's description of the male testicles, attributing to the ovaries the role of a sort of counterweight which allows the passage necessary for the emission of female sperm to open.[56]

The difficulty of reconciling the medical tradition with the demands of Aristotelianism explains the vacillations of Giles of Rome, his constant equivocations and repetitions. As will be seen in the following pages, the perplexity of Albert the Great was at least as great. Nonetheless the solutions adopted by Giles of Rome deserve to be pondered: to make up for the deficiencies of Aristotelianism in the medical domain, he used the *Colliget* of Averroes, and more often than not left Avicenna out of the picture. In this way sexuality and embryology adopted language much closer to philosophy than to medicine and the arguments used were not without their consequences for the new way of representing women. Female sperm was still a product of inferior quality, whose role was negligible. It could at most equal the second male secretion – the prostatic humour. This physiological detail, similar in this respect to the Aristotelian system as a whole, turned woman into an inferior being, subordinate to man. One finds in the text the expression, whose use had become widespread, of *mas occasionatus* (a male whose purpose has been thwarted). The emission of female sperm was granted no importance in the text of Giles of Rome, and was seen as of hardly any significance in a way of thinking completely preoccupied by the question of reproduction. If little attention was paid to the emission of female sperm, then all the pleasure had to be caused by the reception of the male seed. Without casting doubt on the intentions of Giles of Rome, one might suggest that he supplied arguments capable of clearing the male of all responsibility in the woman's quest for pleasure. Thus ideas that were closest to those of modern physiology, by denying the existence of a real female 'sperm', are far less powerful tools for psychological investigation and understanding than the ideas expressed by scholars before the rediscovery of Aristotle.

One cannot but be struck by another conclusion on reading this treatise: the irresponsibility of the woman. Her body lies outside her control; she is mere fertility. She may become pregnant without experiencing orgasm, without any pleasure – practically without knowing it. She is thus the perfect antithesis of the male, who acts on this unconscious mechanism in a responsible and conscious way. Plato had seen woman as nothing but womb: in the Middle Ages, she was still a creature who had something incomprehensible about her.

The same desire to reconcile medical sources with the Aristotelian tradition can be found in Albert the Great. More 'naturalist' than Giles of Rome, the Universal Doctor is not content merely to exercise his powers of thought on the different texts at his disposal: he attempts, at the price of often unsurmountable difficulties, to fit the data of medical or more simply human experience into the theoretical framework. We must leave out of our account the *Questions on Animals*, the echo, as recorded by Conrad of Austria, of courses taught by Albert the Great in Cologne in 1258. These *Questions* hardly diverge from Aristotelian theory and propose solutions similar to the conclusions adopted by Giles of Rome. Far more rewarding is a reading of the *De Animalibus*, a vast paraphrase of all of Aristotle's zoological works. The genre chosen, which hovers between commentary, compilation and quotation from the original treatise, gives him greater liberty than the framework of the scholastic question, that symbol of medieval formalism. The *De Animalibus* reflects the ins and outs, the advances and the dead-ends of real research; basing his arguments on the different authors he reads and quotes, Albert the Great allows contradictions and irreconcilable differences of opinion to appear.

From the many discussions of the question of female sperm, we will concentrate on the attention paid to the different types of humours that may play a part. Woman is subject to different discharges which physiology explains by the excess of cold quality in her, which makes a complete coction impossible. Albert the Great attempts to distinguish, from among discharges other than menstrual blood, those that may be propitious to conception. The link generally established between pleasure and fecundity does not seem to constitute a sufficient explanation: many 'experienced' women have told him that they have conceived without having an orgasm.[57] Albert then turns his attention to the cases in which emission occurs outside the performance of the sexual act. In the first place come erotic dreams, reference to which forms part of a long tradition. For these dreams to announce a future conception, they must take place after the woman's period[58] and they must cause the discharge of a 'ripe, white and viscous' humour, and not one that is 'watery, half-way between water and sperm'. The dream is not considered to be the cause, but the sign of the 'descending of the humour': it is the emission that causes pleasure and it can take place even if pleasure is not deliberately sought. For there to be a subsequent pregnancy, it is enough for the womb to have drawn humour into itself after emission, and for it to retain this humour until the moment of conception. Conception may thus occur without there being any new emission, and thus in the absence of pleasure.

The problem of female involuntary emission, at night or during the day, is tackled several times by Albert the Great. In the *De Temperantia*, he cites the case of nuns who told him that they had never experienced or even imagined any erotic stimulation at the time this emission took place.[59] Likewise, one of the *Questions on Animals* gives the example of the enclosed nuns, often affected by involuntary emission without their imagination having had anything to do with it.[60] Albert's experience as a confessor backs up his medical knowledge, and he imputes these phenomena to the sole functioning of the 'expulsive virtue'. One can thus suppose that leucorrhoeas of functional origin[61] were also included under the name of 'female sperm', whether these were the discharges that take place in a girl in the absence of ovulation, or the leucorrhoeas that sometimes accompany the formation of the cervical glair in the ovular phase. In the latter case, modern physiology shows that it is justifiable to predict a possible conception. More frequently than not, all non-menstrual 'defilement' was thought of as the result of pleasure, and necessarily made women – especially when they were sworn to chastity – feel guilty enough to reveal such things to their confessor. It required all Albert's intellectual honesty for the Universal Doctor not to add an unequivocally negative interpretation or some misogynistic comment.

From a theoretical point of view, except where he is merely paraphrasing the texts of Avicenna or Galen without adding any commentary, Albert the Great maintains a moderate position. The female seed merely helps in reproduction, and is nothing but a material substance. To the objection that it merely duplicates the menstrual blood, he replies by sharing out, between the two female substances, functions that Aristotle ascribed to the male and the female respectively; but he limits them to the realm of material substance: 'Reproduction happens materially from what is called *sperma mulieris* and the nourishment (of the embryo) is ensured by the menstrual blood.'[62] It is clearly the term 'sperm' that throws Albert into the greatest perplexity, as it does most medieval authors, even the Galenists. The *De Secretis Mulierum*, sometimes ascribed to the Universal Doctor – without any very high degree of probability – uses *menstruum* both for menstrual blood and for the other humour.[63] Pietro d'Abano uses conjointly 'what is called female sperm', 'white moisture' and 'drop'. The answer to the question he devotes to the subject in the *Conciliator*, furthermore, does not go as far as the position of Albert the Great: the ovaries merely serve to produce 'a moisture which foments the desire to receive the male seed in the most adequate way'.[64] If the rediscovery of Aristotle marked a regression in the understanding of the role played by the woman, it nonetheless, thanks to the important debates that followed it, showed up the inadequacy of the word 'sperm'. Three different roles could be ascribed to this improperly named substance, depending on who was writing – or sometimes even in the work of one and the same writer: it could play a part in conception by transmitting the maternal characteristics; it could enable the male seed to be better received; and it could act as a sign of the woman's pleasure. Three processes were thus confused: ovulation, cervical secretion and vaginal lubrication. Albert's evident attempts to understand the situation show that for a mind desiring to reconcile experience and rational explanation, the Aristotelian solution appeared over-simplistic, while the Galenist position, even apart from the philosophical obstacle that it represented, could only propose equally inadequate physiological models.

While the intensity of the philosophical debate died down after the middle of the fourteenth century, the doctors of the end of the Middle Ages opted generally for a cautious Galenism. For example, in the middle of the fifteenth century, Michael Savonarola, having stated that only the man has any part to play in imposing form, omitted to say, in his description of the process of conception, where the 'informative virtue' came from:

The sperm of the man contributes in the first place to fertilization. Must the woman's sperm flow at the same time? Speaking as a doctor, I would say yes, since both contribute materially to fertilization, but the male sperm contributes to it formally as well. And I am not going to take sides in this great disagreement between philosophers and doctors, as much because it is irrelevant to our present purposes as because I do not claim to have any other opinion than that of the great doctor, Galen, and his successors. . . . So the two sperms unite, having been ejected at the same time. . . . They are acted on by the informative virtue which modifies, digests and reinforces the heat thanks to the generative spirit.[65]

The theory of the female seed was to survive long after the Middle Ages. The casuistry of the fourteenth century founded, on an assertion as imprecise as this, a distinction between the sinful act and the permissible act.[66] The sexual act that led to an emission in the woman was considered as illicit. The casuists here wielded a weapon that was particularly efficient in repressing female pleasure because of its very lack of a basis in physiological reality. The idea of female sperm was not always abused in this way, and we will leave it to Descartes, that developer and renovator of the metaphors of Avicenna, to evoke, with a certain lyricism, this mysterious secretion:

I leave the shape and arrangement of the particles of the seminal material quite unspecified: it is enough for me to state that the seed of plants, being hard and solid, may have its parts arranged and situated in a precise way that cannot be altered without destroying their efficacy. But it is quite different in the case of the seminal material of animals, which is very fluid and is ordinarily produced by the copulation of the two sexes. This material is apparently just a disorganised mixture of two fluids which act on each other as a kind of yeast, generating mutual heat. Some of the particles thus require as much agitation as fire has, and expand and press on other particles, thereby putting them little by little into the state required for the formation of the parts of the body.

To achieve this result the two fluids in qustion do not need to be very different. We may observe how old dough makes new dough swell, or how the scum formed on beer is able to serve as yeast for another brew; and in the same way it is easy enough to accept that the seminal material of each sex functions as a yeast to that of the other, when the two fluids are mixed together.[67]

One could hardly have dreamt of a finer rehabilitation of the female seed.

IMPURE MATTER

While research into the nature of female sperm had to follow the tortuous paths of scholastic thought, the inquiry into menstrual blood proved to be much easier. As early as Isidore of Seville, the menses were linked to the lunar cycle by their etymology (*nam luna 'mene' dicitur graece*).[68] There were other etymological exercises based on the word *mens-mensis*. As for the expression 'women in their flowers', often employed with a certain relish by writers to refer to menstruating women, it is found in the text ascribed to the famous midwife from Salerno, Trotula. In it, reference is made to the fact that the menstrual flow is responsible for a sort of regulation of the female temperament. Whereas in men the dominant heat is tempered by sweat, in women the excessive moistness is purged by 'the menses that are commonly called *flowers*, for, just as trees do not bear fruit without flowers, likewise women without flowers are unable to fulfil their function of conception'.[69] The vegetable metaphor was present right at the beginning of the treatise: God has given man the strongest and noblest qualities so that he can carry out his duties with respect to woman, by shedding his seed, as it were, into the field entrusted to him.

Apart from the function of drying the temperament that it performs in the woman who has not been fertilized, the menstrual blood has the role, as we have seen, of supplying the embryo with nourishment once the cervix of the womb has closed. This prime function is accepted by the supporters of the existence of female sperm, as well as by their opponents. The famous diagram representing the changes in the uterus of a pregnant woman shows how the duct meant to carry menstrual blood to the embryo diverges from the route via which evacuation normally takes place.[70] This diagram over-simplifies a complex process which 'Alī ibn al-'Abbās summarizes with clarity, after having, as an authentic Galenist, shown the necessity of female sperm; from the union of the two sperms comes the membrane that surrounds the foetus and which will receive the menstrual blood.

When this membrane surrounding the sperm has completely formed, the menstrual blood reaches it by way of the non-beating veins whose orifices are those places called cavities (*cotyledons*); to it come also subtle blood and animal pneuma via the arteries which extend to the womb. These materials seep together into the substance of the membrane, before it hardens completely. For this reason the blood can enter the cavity because of the softness of the membrane. Thus openings and canals form in the membrane. These canals grow continually wider and do not agglutinate, because materials flow continually through them, the sperm ceaselessly drawing the blood by means of the faculty of attraction that it possesses.[71]

The two fundamental substances, one of which keeps its characteristic quality as inert matter, go to constitute the members, in accordance with a division based on the analogy of colours:

From the sperm itself are formed the white parts of the body, that is, the brain, the bones, the cartilages, the nerves, the membranes, the ligaments, the veins and the arteries, whereas from the menstrual blood are formed the liver and the other fleshy parts, with the exception of the heart, for this is formed from the blood of the arteries.[72]

After the formation of the liver, a vein coming from the consolidated and vascularized chorion brings to it the menstrual blood so as to ensure the nourishment of the foetus up until the moment of birth. If we continue our reading of the work of 'Ali ibn al-'Abbās, we discover that the description of the breasts comes quite naturally after that of the uterus in the pregnant woman: 'Since the child has just been nourished on menstrual blood, it needs a nourishment close in nature to menstrual blood, and the matter that has this quality is milk, because milk is formed from menstrual blood.' As the transformation can occur only as the result of a powerful coction, the breasts, whose substance is made of a substance similar to milk, are located near the heart which is the source of natural heat. The kinship established between menstrual blood and milk is not without its consequences for the woman's sexual life. Coitus appears particularly harmful during lactation; Avicenna recalls that its effect is to stir up the menstrual blood and to make the milk smell putrid. But what is especially emphasized is the incompatibility between being pregnant and suckling; as Galen emphasized, pregnancy risks affecting the health of the child at the breast. Menstrual blood cannot simultaneously ensure the nourishment of an embryo and turn itself into milk; the simultaneous performance of these functions would thus endanger the lives of two children.[73]

Outside the conditions of reproduction, menstruation is a simple phenomenon of purification, meant to expel the residues that cannot be transformed by coction because of the woman's lack of heat. The belief that this phenomenon only occurred in the human species never ceased to intrigue, and the question 'why do the brute beasts not experience the phenomenon of menstruation?' was insistently and repeatedly raised. Several reasons could be advanced, all of which stemmed from the same type of explanation.[74] Thus one reason put forward was that female animals, thanks to a more intense physical activity, had their heat raised and thus managed to consume their superfluous matter without any of it needing to be expelled. The same Salernitan question gave a more philosophical answer: in virtue of the perfection that tended to be realized in the human being, the superfluous matter reached the centre of the human body, that is, the womb, before being excreted, whereas in animals it was spread throughout the whole body in the shape of claws or hair. This type of explanation was the most widespread: animals had neither sweat nor menses, but their waste matter was transformed into horns, fur, or claws. A similar line of reasoning accounted for certain secondary sexual characteristics. According to Aristotle, women had neither haemorrhoids nor nose-bleeds; they had less pronounced veins than those of males, and finer and smoother skin 'because the residue which goes to produce those characteristics in males is in females discharged together with the menstrual fluid'.[75] To these remarks on the differences between the two sexes can be linked the extremely widespread quodlibetical question 'Why do women not have a beard like men?' The commonest explanation drew on the way the pores were constricted: in women, the excess of the cold and moist humours rendered impossible the formation of the vapour which opened the pores and solidified into hair on contact with the air outside.[76] But one can also find the beard and the male distribution of hair being treated as an outlet for superfluous matter; this is the case in the *Dialogue de Placides et Timéo*: 'And exactly as men are hotter than women and for this reason are hairier, in the same way males among the animals are hairier or provided with bigger horns than the females.'[77]

While medieval medicine showed a good knowledge of various gynaecological disorders, such as dysmenorrhoea or amenorrhoea caused by famine,[78] it also lent its authority to the most curious statements and nourished a number of popular beliefs relating to menses. These beliefs have been discussed frequently and this is not the place to go into them again.[79] With the help of Vincent of Beauvais, we will merely pick out those assertions accepted by informed circles in the Middle Ages.[80] So: it was believed, in accordance with a tradition already recorded by Pliny, that menstrual blood prevented cereals from sprouting, and soured the must of grapes; that on contact with it, herbs died, trees lost their fruits, iron was attacked by rust and objects made of bronze went black; and that dogs who had absorbed it contracted rabies. It also had the property of dissolving the glue of bitumen which even iron could not break down.

One also comes across the assertion that a child conceived during the mother's period would have red hair, with all the connotations attached to that colour. Starting from the colour itself, a system of the effects of menstrual blood (red, red hair, rust) organized itself in conformity with the way that the analogical and etymological thought of the Middle

Ages typically operated. In this series of associations, we must include the properly medical explanation of smallpox and of measles. These diseases, most often contracted in childhood, are a demonstration of the effort nature makes to purge the body of the menstrual blood retained in its porous parts during gestation. But that is not, perhaps, the essential thing about the beliefs that started to develop, especially from the thirteenth century onwards. The essential thing, rather, is the system of representation of venereal infections associated with leprosy. The child conceived during a woman's period or during pregnancy could be afflicted with this illness. We will discuss this question in our chapter on pathology.

To return to our subject, which is more strictly physiological, we must mention an association of ideas that generated mythico-religious stories, often told in euphemistic terms, but still recognizable in medieval literature. What we have in mind is the prohibition attached to a menstruating woman, a prohibition that was reinforced by the certainty that her body could transmit the poison that had been formed in her organism. Melusine and her avatars, those maidens of literature plunged up to their waists in water, are evidence of this prohibition and the rituals of purification it made necessary. On the other hand, the woman with the poisonous glance deserves a supplementary explanation.

A whole network of beliefs rested on the assertion that the menstruating woman had a gaze capable of dulling mirrors. In this way, a lasting association was established between the woman in this state and an animal from the medieval bestiary – the basilisk, the mere sight of which was capable of causing death.[81] The basilisk was born from a cock's egg. This reflected the idea of dryness in its extreme form. The egg had to take shape in the organism of a cock aged between five and six, which increased the degree of dryness in it. All the bad superfluous matter from the kidneys and the genital organs, that had not been eliminated because of the lack of humour, enabled an egg to be formed – in the intestine, since there was no womb. As there was no meeting of a male with a female element, the production of an animal of a different species was possible, and since it developed out of poisonous material, a poisonous animal – a snake or a basilisk – was created. The explanations of the way the death-dealing act occurred are varied, but, to simplify, let us say that the *pneuma* emitted by the basilisk modified the quality of the surrounding air; or else the venom could reach the heart through the eyes. When the basilisk's gaze met with a polished surface ('a mirror or shield containing pitch'), the venomous humours were reflected back, and the basilisk was killed. The analogy with the maleficent gaze of a menstruating woman was made widespread by the *De Secretis Mulierum*, which returns to this theme several times.[82] Albertus Magnus, furthermore, had devoted one of his questions on animals to the 'infection [= impregnation] of the eyes caused by the flow of menstrual blood'.[83] He explained the process in this way: the eye being a very passive organ, it receives during the woman's period the menstrual fluid which imbues it; so every object placed in front of a 'menstrual' eye will be infected. In the absence of the notion of contagion, it is the air which, being corrupted on contact with a harmful source, transmits the disease. From the point of view of physiology, this explanation agrees perfectly with the Aristotelian and Galenic theories of vision, in which air plays the role of the necessary intermediary between the eye and the object. So it is not the eye itself, as Albertus Magnus emphasizes, but the noxious vapour that it gives off, which imbues everything placed near it. When the menstruating woman turns her gaze to a reflective surface, she may feel a pain in her eyes or else, under certain conditions, leave on the mirror an indelible stain. In his article, 'Véronique ou l'image vraie', Claude Gaignebet, as well as drawing up a very full list of the implications of this belief, indicates its Aristotelian origin; and in fact, one can read in the *Treatise on Dreams*: 'If a woman looks into a highly polished mirror during the menstrual period, the surface of the mirror becomes clouded with a blood-red colour.'[84]

The synthesizing and circulating of all these elements of the belief seem to have occurred at the end of the thirteenth century. The story of the Venomous Virgin, which was extremely widespread at this period, brought these themes together and lent its exemplary authority to them.[85] One finds, stated explicitly in works of scientific popularization, an idea that was always latent in texts of a higher scientific level: that the female organism was capable of producing poison, in other words death or illness. The period following the menopause evidently made the woman even more dangerous, because she had become incapable of eliminating the superfluous matter from her organism. In *Les Admirables secrets de magie du grand Albert et du petit Albert*, the phenomenon is explained in the following way:

If old women who still have their periods, and certain others who do not have them regularly, look at children lying in the cradle, they transmit to them venom through their glance. . . . One may wonder why old women, who no longer have periods, infect children in this way. It is because the retention of the menses engenders many evil humours, and these women, being old, have almost no natural heat left to consume and control this matter, especially poor women, who live off nothing but coarse meat, which greatly contributes to this phenomenon. These women are more venomous than the others.[86]

This translation of a passage of the *De Secretis Mulierum*[87] shows clearly that woman was venomous by virtue of her very physiological mechanism.

The explanation that enabled as many women as necessary to be identified as witches and sorceresses was already well established as early as the end of the thirteenth century. It is noticeable that poor food was considered to make old women from the lower depths of society even more dangerous. Never had misogyny gone to greater lengths, never had it been furnished with more powerful arguments, since women could now be the principle of destruction of the very species to which they belonged.

The menstruating woman was also the source of perverted offspring: 'If the hair of a menstruating woman be taken, and placed under a dung-heap or clod of earth, or where the dung was made during winter or summer, by the virtue of the sun there will be engendered a long and powerful snake.'[88]

A variant on this theme may be found in a version of the *Secrets des Dames*: 'And whosoever were to take a hair from the pubis of a woman and mix it with menses and then put it in a dung-heap, would at the end of the year find wicked venomous beasts.'[89]

The superstitions that we have mentioned were consequences of the normal functioning of the woman's physiological system. There was no gap between beliefs attributed to ordinary people and the scientific thought of the age. The discourse hostile to woman enjoyed the protection of scientific authority and deployed such well-established arguments that it could indulge in every excess. Even if they themselves did not demonstrate the exuberant imagination that could be found in non-scientific writings, doctors were convinced that the menses were harmful. Another proof was supplied by cases of hysteria, as we will show in our chapter on illness; when the harmful element could not find any outlet, that is, when there was retention of the menstrual fluid before the menopause, the venom turned against the organism that secreted it. Here, we again encounter the idea of the witch. The only possible salvation for women remained motherhood.

Nonetheless, the notion that menses were the raw material of the embryo created its own difficulties because of the very impurity of this residue. It was often this impurity that led the most moderately Galenist writers to grant a certain influence to the female sperm, a substance judged to be 'more beautiful and noble' because better controlled. In his *Commentary on the Sentences*, Thomas Aquinas admits that, in women, the product of the blood is transformed into a material substance susceptible to the action of the male seed. In the *Summa Theologiae*, on the other hand, he establishes an intermediary between the menstrual blood and the embryo. This intermediary is a special blood, 'digested over a longer time', cleansed and purified: 'This purified blood (which remains very pure in the Virgin Mary) is however always stained by a certain corruption, by an impurity due to desire, for it is drawn into the uterus only by copulation.'[90] For Aquinas, menstrual blood is merely the residue left from the formation of this second blood, and it contains nothing but impurities. Saint Thomas thus admits that there are three humours: a semen which does not participate in the process of generation, but facilitates the union of the partners; menstrual blood; and the blood that goes to make up the embryo, which is formed from this impure blood. There is hardly any doubt but that the theory accepted by Saint Thomas enables him to keep the residue and the impurities of the menses away from the embryo. Lastly, it gives him the possibility of dividing menstrual blood into two parts, so that the embryo will not be sullied by an influx of impure blood. His thoughts on Christ's conception make it clear what theological reasons motivated this scientific choice.

The distinction drawn between several qualities of menstrual blood can furthermore be found in the medical sources. The *Canon* of Avicenna, having given an account of the Aristotelian theory of coagulation, returns to a more Galenist position:

After that, the blood which is voided by the woman at the time of menses is used for nourishment. Part is transformed in accordance with its similarity to the spermatic substance: it forms the parts of the body that have come from the sperm, and it increases the sperm by nourishing it. Another part does not act as nourishment but, through coagulation, is used to fill the empty spaces of the principal parts of the body, and makes flesh and fat. A last part consists of superfluous matter and is no good for either of the above-mentioned purposes: it stays put until childbirth, when nature expels it as being superfluous.[91]

Indeed, some account had to be given of the liquids expelled by women in childbirth; their expulsion necessarily signified that they were not only useless, but impure. It was on this type of distinction that Thomas Aquinas based his arguments, while refraining from granting to the more noble substance the role ascribed to it by Avicenna. The ambivalence of the menstrual blood, both venomous and fertile, raised as many questions for doctors as for theologians. An original idea was suggested in twelfth-century Muslim Spain by ibn-Zuhr or Avenzoar whose *Kitāb at Taysīr* was used in its Latin translation only at the end of the Middle Ages. According to Avenzoar, people must shake off the notion that the embryo is nourished in the womb by the menstrual blood; this false notion stems, in his opinion, from a peculiar feature of Galen's vocabulary:

The embryo is nourished by blood descending to the vulva. As Galen called the said blood 'menstrual', certain scholars have thought that the embryo was nourished by the menstrual blood, but this is not the case. Galen was merely following the usage of the Greeks who call 'menstrual' all blood which descends to the vulva, just as they call 'quinsy' every abscess occurring in the throat, whether it be sanguine, choleric, phlegmatic or melancholic. . . . It must be considered certain that if the embryo were nourished on the same blood as that which is expelled by women, it would be completely unable to live. In reality the embryo is nourished by most excellent blood.[92]

While it does not take its nourishment from it, the embryo is, nonetheless, during gestation, soaked in blood usually expelled in the menses, just like dough which continues to rise even after the vessel containing it has been washed clean of any leaven. It is impossible for menstrual blood to play any part in nourishment – a fact demonstrated by the problems it causes in a woman who is not sufficiently purged at the time of childbirth, and in a child 'saturated' with it during pregnancy. The harmful liquid which has soaked into the body of the foetus mixes with the humours when the newborn child is being fed and can only be eliminated at the cost of an effort comparable to that made in extracting butter from milk. Avenzoar's whole demonstration is meant to account for the aetiology of smallpox and measles. As physiological processes were conceived in terms of nourishment and digestion, the impure quality of menstrual blood could only lead scholars to distinguish between the different states of the blood. Thanks to the systematic weighing-up of contradictory opinions, medieval science clearly perceived the inadequacy of the models inherited from antiquity in accounting for the complex phenomena of reproduction; the role played by menstrual blood raised as many problems as that of 'female sperm'. Avenzoar's text emphasizes too the dangers of an over-simplification present, for example, in works of popularization; scientific theories cannot be maintained unless they show a logical refinement which appears only rarely in texts whose main purpose is of a different order. The example of menstrual blood illustrates perfectly how one can move from one level of culture to another – a move thanks to which science may itself generate beliefs which it then assimilates back into itself and justifies.

HIGH SPIRITS

The medical definitions of coitus showed, as one might have expected, the importance of the purposes of reproduction. We will mention, from among other similar definitions, that given by an anonymous thirteenth-century author: 'Coitus is the mutual action which man and woman perform by means of natural instruments in order to perpetuate the species.'[93] But the author adds immediately: 'Almost all men desire coitus because of the pleasure it gives, and few because they hope to beget children.' Although occupying a central place in the definition, reproduction no longer plays any part in medical discussions, except as one of the consequences: coitus remains, above all, one of the factors governing sickness and health. One must thus envisage a threefold teleology: from a naturalist point of view, the purpose of coitus is the preservation of the species; for doctors, it is the preservation of health; and for the individual, it is pleasure. The role of the therapist consists in maintaining a balance between these three demands and in analysing mutual obstructions or incompatibilities between them.

When doctors are not specifically discussing either the organs proper to each sex or the preconditions of reproduction, but the physiological mechanisms that come into play during the act, they refer essentially to the way the male functions. Medieval authors follow an invariable model, and repeat what we have read in William of Conches. Three elements are necessary; heat which vivifies and dissolves, the spirit which gives the impetus, and the humour which is dissolved. The analyses of anatomy and physiology that we have presented leave hardly any doubt on the functioning thus schematically described. The second element that plays a part in the act did however give rise to an ambiguity: the anonymous author that we referred to above mentions 'the spirit or flatulence'. This ambiguity in vocabulary is present in most medieval authors from the thirteenth century onwards, without ever being completely resolved. While they state that the process of reproduction brings into action a spirit or *pneuma*, the term 'flatulence' often seems to them more adequate as a means of describing the mechanism of erection and ejaculation. Thus, Albert the Great in the *De Animalibus* says:

For one must not forget that in every ejection of sperm, flatulence is involved; the ejection occurs in several movements, for the flatulence ejects first one part, then another. The situation is the same as in vomiting, which is caused by the force of what is vomited: the number of these spasms of vomiting corresponds to the number of climaxes experienced in coitus.[94]

The term 'flatulence' suggests, rather than a noble vapour like spirit, a by-product, a sort of gas whose nature is not clearly defined. Furthermore, this lack of precision was already present in the *Canon* of Avicenna: 'Erection is due to a strong flatulence brought on by the *spiritus desiderativus*.'[95] This divergence of vocabulary no doubt demonstrates that a somewhat different idea is at work and the prescriptions which, inherited from antiquity, ascribe great importance to the intake of flatulent food in the arousal of desire, could only be reinforced by this.[96]

At the same time as being the motor of sexual appetite, this vapour contributes greatly to the pleasure experienced, as Galen says:

Now if I may in order to make my narrative clear give little, insignificant examples of the wonderful works of Nature, reflect with me on the sort of thing that happens when these serous humors are heated, as they frequently are, especially when acrid humors collect under the skin of the animal and then itch and make it scratch and enjoy the scratching. Whenever, therefore, there is not only a moisture of this sort needing to be evacuated and hence stimulating and provoking its own evacuation, but also a great deal of warm pneuma requiring to be exhaled, we must consider that there is an extraordinary, excessive pleasure. And since Nature has made these parts much more sensitive than the skin for the sake of the same usefulness, it is no longer to be wondered at that the pleasure inherent in the parts there and the desire that precedes it are more vehement.[97]

Following Plato and Aristotle, Galen justified the existence of this pleasure by its biological purpose: nature, knowing that its creatures are not endowed with perfect wisdom, has replaced this wisdom with an all-powerful stimulation so as to ensure the preservation of the species. All medieval writers repeat this type of justification, generally adding that the enjoyment helps to overcome the disgust caused by having to use such 'shameful' organs.[98] This reminder of the biological purpose of pleasure and the reference to its usefulness at the moment of conception, at least in the opinion of the supporters of female sperm, could only tally with the demands of Christianity, for they implied the recognition of limits not to be overstepped. When he attempts to define the virtue of temperance, Albert the Great establishes a parallel between two forms of enjoyment linked to the sense of touch: food, the aim of which is the preservation of the individual; and the pleasures of love, the aim of which is the preservation of the species. One of the questions raised is the following; why should one apply a stricter discipline to the latter?[99] The answer brings in a distinction between two types of regulations: in the case of food, the limits come from outside the individual, whereas in the case of the pleasure linked to reproduction, it is the organism itself that lays down the conditions. For the aim and purpose of coitus to be fulfilled, the excess product supplied to this end must be digested and absorbed in the appropriate place. Over-vigorous coitus leads to an emission that is incapable of causing conception: the timing and the intensity must be submitted to a set of rules. Likewise: 'Determinatur EX LEGE quod nullus accedat ad non-suam'. 'It is laid down BY LAW that no man may have intercourse with any woman except his wife'. For indeed, if just any man went with just any woman, or several men with one woman, conception would be prevented by the 'slippery nature' of the womb, as can be seen in prostitutes who are sterile. We here come across an idea already expressed in William of Conches and it is clear that in Albert the Great, physiology is a justification for ethics. Sexual intemperance is not merely a sin, but a vice from the point of view of nature, whose designs are thereby thwarted.[100] On the other hand, there is no law governing the more 'innocent' forms of sexual pleasure, such as caressing the breasts or kissing: these activities do not concern any principal virtue.[101] Since the only misdemeanour is thwarting the function of reproduction, how then can the maintenance of virginity be justified from the point of view of the naturalist? Albert the Great replies that reproduction is not always necessary; in a time when the human race is multiplying sufficiently rapidly, virginity can be praised for its beauty and purity, since it involves no inconvenience or danger for the species. From the biological point of view, the non-fulfilment of a function seems less detrimental than using it for other purposes. If Albert's opinion can be fitted perfectly easily into a naturalist framework, it does not entirely agree, as we shall see, with the point of view of the doctor who links the sexual act to processes other than that of reproduction.

A first approach to the psychological aspect of pleasure is made by way of a consideration of the differences that may be observed between the sexes and between one species and another. The traditional question: 'Why does woman, although she is of a colder and moister nature than man, feel a more burning desire?'[102] at first receives an answer linked to physics: damp wood takes a long time to catch fire, but it burns for longer. The quality of cold seeks for its opposite, and so the womb rejoices when it takes in hot sperm . . . like snakes who, in their search for warmth, slip into sleepers' mouths. By opting in favour of the existence of female sperm, the *Pantegni* spread another kind of answer: woman enjoys both emitting her own seed and receiving male sperm.[103] The arrival of Aristotelianism was to suggest a completely different interpretation. Woman's excess of moistness, her immoderate lust and her passivity, made of her a creature *semper parata ad coitum*, who once the act had been completed remained *lassata sed non satiata*, as Juvenal put it.[104] From the thirteenth century onwards, the answer to the traditional question quoted above was thus given a new twist: women experienced a pleasure that was greater in quantity, but lesser in quality and intensity, than men's. The reasons put forward were not restricted to the interaction of opposite qualities, but lay outside the field of physics: instability and lack of satisfaction were consequences of the weakness that affected the female faculty of judgement, characterized by its 'crookedness'.[105]

The differences between coitus in animals and in humans also led writers to define the mental processes and the emotions that precede and accompany the physiological mechanisms. The realization that women, as opposed to female animals, continued to want sexual relations even after having been impregnated, was such as to constitute a setback for the idea of biological teleology.[106] The principal argument put forward to account for this human characteristic granted an important role to memory. In the human being, desire was aroused not only by the natural appetite, but by the animal appetite (that is, the appetite born of the soul). The sensations of former experiences were retained in the *imaginatio*, and memory, by recalling them, encouraged one to renew the pleasure. That left the differences which affected human emotions to be explained. This was the aim of the astonishing question, raised for instance by Albert the Great: 'Why do all animals, with the exception of humans, make a noise during coitus?'[107] Indeed, the cock and the horse cry out, dance or 'sing', whereas humans 'arm themselves in silence and in secrecy as if going to war'. The silence ascribed to the human race demonstrates too the sense of shame that characterizes it, and the attraction that secrecy has for it (*quanto occultius, tanto dulcius*). But it is also the accidents of the soul, the passions, with their hybrid status as both physical and mental states, which show up the difference. Under the influence of intense pleasure, the heart of a man or woman contracts and, just as when they are afraid, they are struck dumb. Furthermore, whereas heat and the movement of blood make people bold in their desires, the congealing of spirits gives birth to pusillanimity once the act has been completed. These phenomena serve to account for the behaviour of the adulterous man:

Before coitus he is not afraid to go after the wife of another man, even in the latter's presence, for he is subject to a powerful motion of the spirits; but after the act, because of the debilitation that he suffers, he is afraid and runs away at the mere sight of a rat or a cat.

The way that note was made of the mental states and emotions aroused by desire for, and performance of, the sexual act led medieval authors to integrate, more subtly than had been done in antiquity, the psychological dimension into their depiction of physiological mechanisms. It is also noticeable that, generally speaking, being constrained to exclude the soul from their preoccupations, they were led to adopt the theories that were the most capable of giving a physical explanation of psychological phenomena. But it was nonetheless only at the cost of having to concentrate to a sometimes exaggerated extent on the physical body that they were enabled to tackle psychology. Right from the Salernitan period, doctors applied themselves to defining the essence of sexual pleasure; in their accounts, we again find that the eyes were given great importance. This importance was frequently pointed out:

Love is nothing other than a pleasure accompanied by joy and all pleasure rises either from the outside, that is, from the soul, or else from the inside, that is, from nature. Each of these two types of pleasure operates thanks to the appropriate instruments and by means of certain parts of the body, namely the eyes. The spirit transmitted by the optic nerve is emitted outwards to apprehend the things outside; having grasped those things, it embraces them and represents them to the superior part of the soul.[108]

The remembrance of the thing perceived is imprinted on the memory and imagining a renewal of this joy leads to an emotional process that sets the physiological mechanisms into motion:

And thus the psychical virtue arouses the natural virtues to put their activities into operation. When a pure blood is engendered in the liver, and floods, nourishes and reinforces the members, nature feels pleasure thereat as if under the influence of a similar and benign nourishment, since human nature has the same complexion as blood, that is, hot and moist. The natural heat is thus aroused by the psychic virtue, and by their combined action, the blood contained in the liver moves and in moving emits heat; from it there evaporates a smoky cloud which, when it has been made subtle, spreads from the liver to the heart. From the heart the spirit moves to the penis by means of the arteries, and makes it stiffen.

The psychic virtue, transmitted by the *pneuma* of the same name, thus gives an impetus to the three elements necessary for the performance of the sexual act: heat, moistness and spirit. The thing to be remembered from this Salernitan question is the psychophysiological explanation; it lays down the main lines of something developed more subtly in later works, thanks to the contributions of Aristotle and Avicenna. The links between psychology and sexual behaviour are most frequently tackled from their negative angles. The treatise of Ishâq ibn-'Imrân, translated in the eleventh century by Constantine the African under the title *De Melancholia*, spread two ideas that were frequently developed: on the one hand, too rigorously an ascetic life risks causing the illness of melancholy; on the other hand, this illness can find relief thanks to coitus.[109] Likewise, the *Canon* of Avicenna cites among the benefits of the sexual act; the 'expulsion' of a dominant train of thought or of an obsession, the acquisition of boldness, the control of excessive anger and, of course, the dissolution of the spermatic vapours that accumulate in the brain of melancholics.[110]

The originality of the Middle Ages lay in their bringing together the facts provided by psychology and physiology, so as to constitute a model capable of accounting in detail for the overlapping of physical and mental states. One of the finest examples of this is provided by the treatment of the illness called 'heroic love'. Obsessional love came within the purview of pathology as early as antiquity, and was the subject of extensive commentaries thanks to the adopting of the philosophico-medical theory of the internal senses.[111] In the shape in which the Christian West took it over from the Arabs, this theory located inside the three cerebral 'ventricles' the seat of forces or virtues which, transmitted by the spirits, played the role of intermediaries between the five external senses and the intellective powers of the soul. In the treatise which Arnald of Villanova devotes to heroic love, we again find the different milestones in the development of the Salernitan question on the nature of pleasure. The disease (close to melancholy) which is called *amor hereos* or *amor heroicus* comes into being in the same way as any 'normal' feeling.[112] On perception of the object, pleasure is felt; if the estimative virtue, placed in the median ventricle of the brain, judges this pleasure to be very great, the imaginative virtue and the memory, situated respectively in the anterior and posterior ventricles, retain in all their force the impressions received, and also the intentions of the estimative virtue. Pathological obsession and derangement occur when this last faculty estimates that the pleasure to be obtained surpasses all others and constitutes the one sole good to be sought. For Arnald of Villanova and the majority of doctors, responsibility for this disorder lies with the judgement, whose aberration results from the intensity of love. The heart, the seat of the passions, is the prime mover of the derangement: at the sight of the lovable being, the vital spirits are inflamed and stirred up, and confuse the judgement, when they reach the median ventricle of the brain; by proximity, they also dry out the anterior ventricle, something that leads to the fixation of the impressions received and to obsession. The preconditions of the illness are all fulfilled if the passion of love is thwarted, that is, if the carrying out of the sexual act is not the natural conclusion of the psychological mechanism. The break in the sequence of psychophysiological processes leads to grave disorders, both physical and mental. Shadows under the eyes, a yellow hue on the face, thinness and an accelerated pulse, together with an excitement shown by rash words and gestures, make up the picture of the 'heroic' lover. Dominated by his exclusive passion, he loses his liberty and risks falling into the supreme aberration of madness, or of neglecting to carry out his vital functions. To counteract this dangerous illness, doctors proposed a form of therapy most of the details of which were taken from the Ovidian *Remedies for Love*. The cure must first of all take advantage of all opportunities for distraction, so as to drive away the image of the loved being. One of the most efficient modes of action consists of transforming desire into its opposite, namely repulsion or disgust. Ovid recommended calling on a woman before she has performed her toilet, or deliberately gazing at soiled sheets. Avicenna advised that the loved one should be constantly disparaged – a function that old women could fulfil with great success, by making out that the chosen woman was the most repulsive person imaginable. Medieval doctors took up this advice, adding their own exaggerated colouring, sometimes going as far as obscenity.[113] If nausea appears to have been one of the antidotes to love, the sexual act constituted its natural outlet. So doctors insistently recommended it; ideally, it should be performed with the loved person, but since the illness most often resulted from the very fact that this was impossible, it was recommended that the act be performed with different people, so as to avoid the risk of a new 'fixation' of the passions. Morality retired into the background when it was necessary to re-establish the temporarily interrupted sequence of vital processes.

In the *Speculum Medicinae*, Arnald of Villanova proposes a subtler analysis of the effect which the faculty located in the median ventricle of the brain has on the body.[114] It can influence the state of members or organs in two ways: either indirectly, via the passions – and thus via the heart – or directly, as when it judges that an object is pleasant or harmful. This last mode of action finds its realization in two similar functions; after a mental inclination has led one to pursue a particular pleasure, the reproductive organs prepare for coitus; likewise, at the sight of something repugnant, nausea sets in. We again find the analogy between vomiting and coitus; in Albert the Great it served to account for the phenomenon of expulsion following a spasm, whereas Arnald of Villanova uses it, in a more philosophical context, to explain the physical effects of two sensations of the same degree, although opposite in quality – disgust and desire. The estimative virtue, which we will call judgement for simplicity's sake, operates not only at the time of the decision, but during the progress of the act. This realization supports the ancient belief, according to which the images that present themselves to the spirit in the course of conception influence the form of the embryo. In connection with this, a Salernitan question cites the case of adulterous relationships which can give birth to a child resembling the deceived husband and not the natural father; the explanation is that shame brings to mind during the act the image of the absent man.[115] Cautiously ascribing the first cause of this phenomenon to God's power, Arnald of Villanova summarizes everything the doctor must know: the spirit or *pneuma* is a receptive agent and transporter of forms which, although deprived of sensitivity, creates, like air on a mirror, a perceptible impression on the parts of the body concerned. The *species* thus carried by the spirit modify the design laid down in the informative virtue contained in the sperm.

At the same time as fitting easily into an extremely important philosophical, and even Averroist, framework, the question of 'heroic' passion stimulated medieval thought and helped it to work out the links between mental states and physiological mechanisms.[116] Judgement, placed under the rule of reason, remained sovereign when it came to determining whether sexual pleasure constituted a good to be sought; but as soon as it set out spontaneously after pleasure, or else was constrained to do so by the force of passion, the physiological sequence was irreversible and any obstacle risked creating problems for one's mental and physical health. While it represented a hindrance to freedom in the eyes of the philosopher or the theologian,[117] human love was also treated with suspicion by the doctor, if it did not become a practical reality involving the natural functions. Furthermore, taking the inter- ferences between psychology and physiology into consideration helped to reinforce the medical 'regulation' of sexuality: if too frequent coitus led to a drying up and consequently to a vacuum, especially in the 'prow' of the brain where impressions provided by the senses were grasped, excessive abstinence led to heaviness, depression, and even a deranged mind. The virtue of continence, judged acceptable by the naturalist when the preservation of the species was ensured, risked coming into conflict with the demands of medicine, orientated as they essentially were towards the preservation of the individual. Thus supplied with an explanatory system which enabled it to link soul and body, could medieval medicine also find room for an art of love?

3

Medicine and the Art of Love

The fact that a physiology of pleasure came to be included in the description of bodily mechanisms invites us to determine exactly what was at stake in medical discourse, and to locate it somewhere between the processes of repression and the products of erotic literature. Did medicine really produce only a normative discourse, subordinate to the demands of morality,[1] or else, in those centuries when oriental influences and Christian values interpenetrated, was it able to speak another language?

THE QUEST FOR MASTERY OF SEXUALITY

The penitentials give us some quite important information on the sexual practices of the high Middle Ages. The investigations conducted over the past few years into this type of writing mean that we need mention them only very briefly.[2] They were in use from the sixth to the eleventh century. From the start they linked together, as J.T. Noonan remarks, contraception and magical practices:

This review of the texts not only serves to establish the close association between abortion and magic, between contraception and magic, and between poison, abortion and contraception; it may also suggest that with the authors of the penitentials, as with Caesarius of Arles, opposition to magic reinforces the opposition to contraception. Contraception is not condemned merely as magical, but the hostility to it has added force from the orthodox Christian distrust of pagan magic.[3]

Knowledge of love philtres and sterility inducing potions, and recourse to abortion, all belonged to the diabolic world, and the eventual fate of the outsiders who indulged in these activities could already be guessed at. Herbal potions were in favour as early as the first centuries of the Middle Ages, and Salic law was already expressing its opposition to these practices. One has to consider the protective role that laws or precepts of this type may have played. Local contraceptives are capable, in certain populations in the Third World, of causing alarming sterility rates. What exactly is meant by the phrase 'a potion that has caused definitive sterility?' In how many cases did the recipes suggested cause serious gynaecological accidents? On this point, law and religious precepts were in agreement – which could be evidence for an unconscious desire to safeguard the fertility of the population. It was this aim, over and above moral or religious justifications, that was perhaps the true role of the regulations laid down by the penitentials – a hypothesis that we formulate with all due reserve.

The doctrine of the penitentials is simple: reproduction is women's principal role and any emission '*extra vas*' is strongly condemned, especially if it involves oral or anal practices. The essential source of information as far as we are concerned is in the different penances laid down and, as has been said, they create a false impression of severity. Other considerations mitigate their rigour: punishments involving fasting for a very long period may be commuted into a shorter period of prayers. As far as abortion is concerned, the theme of the *paupercula*, the poor woman, is frequently found in these writings: the penalty is in this case reduced by half of that laid down for a woman enjoying a normal economic situation. As for coitus *a tergo*, it is condemned, but with only relative severity in the penitential of Burchard of Worms (between five and forty days' fasting). No medical reason is advanced by the monks: they implicitly observe natural law.[4]

All of this interplay of prohibitions was aimed at making a vigorous, natural sexual life the norm – one, that is, that was forced to conceal itself far less than was the case in the following centuries. One must not forget, either, the situation of the authors of these texts; it may explain the severe punishment that they meted out to certain forms of wrong-doing. Communities were no doubt constantly obliged to suppress homosexual practices.[5] Indulgence was the rule when adolescents (*pueri*) were involved; coitus *inter femora* practised between men was relatively lightly punished (one year's fasting), whereas homosexuality was punished more severely (ten years). It would also seem that a contrast was established between passive homosexuality (*mollities*) and active homosexuality. Summarizing these tendencies in crude and simple terms, one may say that the penitentials are evidence of the obsession that haunted the Middle Ages from beginning to end and which underlay the demand that the sexes be rigorously separated. This demand evidently corresponded to the presuppositions that held sway in anatomical description.

One could also say that these texts aiming at repression constituted an inverse art of erotic information; sexuality was talked about, and as a consequence there was the formidable problem that the penitent might inadvertently be told something new – and thus be tempted to sin. We will see that this kind of argument was put forward by Avicenna's commentators, who imposed a kind of self-censorship, refusing to interpret or gloss certain passages of the *Canon*. Very soon, ecclesiastics were faced with the same problem. Noonan points out that Theodulphus, Bishop of Orleans at the beginning of the ninth century, 'instructs his priest to inquire in confession about fornication, a sin which includes coitus interruptus in his legislation'. Burchard of Worms advised that women should be questioned 'mildly and gently' about every kind of sexual behaviour: female homosexuality, masturbation, bestiality, incest, abortion and contraception.

An important stage in the history of medieval sexuality was reached when, from the eleventh century onwards, a ban on contraception was established in a pontifical collection. The condemnation of contraception, as formulated by Saint Augustine, was repeated by the two great authorities of the Middle Ages on religious law: Gratian and Peter Lombard. One finds in them a firm condemnation of all behaviour that excludes insemination; Gratian in the Canon *Adulterii Malum* establishes a classification of different sins:

The evil of adultery [*Adulterii malum*] surpasses fornication, but is surpassed by incest; for it is worse to sleep with one's mother than with another man's wife. But the worst of all these things is what is done contrary to nature, as when a man wishes to use a member of his wife not conceded for this.[6]

Over and above these constant reminders of the ban on unnatural, that is to say sterile, orifices, some authors took a more hard-line attitude to contraception than others. For Peter Cantor, health reasons did not justify using a contraceptive, an opinion that went against the advice of doctors. In 1230, the Dominican Raymond of Pennafort composed a collection of *Decretals* which became Church law for more than six hundred years. Contraception was condemned. In medical treatises, however, the long list of potions and pessaries continued to appear.

The received opinion is that the Church's position hardened, for it found itself faced with adversaries of procreation such as the Bogomils, then the Cathars. Although the sexual behaviour of the members of heretical sects deserves to be investigated in a more detailed way than the limits of our present study allow, it must be recognized that a number of these sects represented tendencies which could have entailed the death of the species. Each time, the Church thought it would have to repeat Saint Augustine's struggle against the Manichaeans all over again. In connection with this, we will be studying the question of courtly ideology in more detail later.

The defining of what was allowed in sexual behaviour was followed – or even superseded – by the desire to establish a norm in conjugal sexuality and to investigate the legitimacy of pleasure even in this context. Gratian had suggested that 'Those who copulate not to procreate offspring but to satisfy lust seem to be not so much spouses as fornicators.' Certain authors took this tendency even further and proposed that even pleasure in the act of procreation was a sin: to act without sin and still have children, the 'holy man' suffered the pleasure, 'just as one eating honey to feed himself might suffer the sweetness of the honey'. This comparison, put forward by William of Auxerre, shows the furthest point reached in the repression of pleasure.

There is, as we have said, a flagrant contradiction between the growing mass of prohibitions laid down by theologians, and the influx of information relating to contraception. From the high Middle Ages up to the fifteenth century, manuscripts presented, either one after the other, or else intercalated between the principal texts, innumerable isolated prescriptions; among these prescriptions, often on the borderline between magic and medicine, it is rare not to find a piece of advice *ut mulier non concipiat*.[7] Sometimes taken from longer works, they were not normally accompanied by the precautions that scholars habitually added; thus, the advice recommended by Trotula when a woman was inclined to try and avoid impregnation because of the narrowness of her organs or because of the fear she might feel as a result of a previous particularly painful delivery, were recopied and translated without the usual discussion. The 'therapeutic' aim of the prescriptions was more often than not effaced by the way they were bluntly presented in the midst of a heterogeneous list of remedies or on some folio left blank in a manuscript.[8] Furthermore, from the thirteenth century onwards, information on contraception was in wide circulation, through even less clandestine channels. It could be found in Aristotle's *History of Animals*, but also in the writings of Rhazes or Avicenna which were put onto the syllabus of the universities. With the work of these two doctors the Middle Ages had at its disposal as much information on contraception as the Greeks or the Romans. A curious situation – this gap between the information people had access to and the demands of Christian morality! The means used were liquid spermicides and pessaries. As for talismans carried around the neck, they were supposed to prevent fertilization. Avicenna, and after him Albert the Great, expressed reserves about the effectiveness of the procedure.[9] It may appear surprising to find information on contraception in the work of the Dominican Albert;

it can be found in his paraphrase of the pseudo-Aristotelian *De Plantis*. Relying on his sources or on direct information, Albert lists, in book VI of his *On Vegetables and Plants*, the substances traditionally used by doctors or by the *incantatores* and *magici* to prevent conception or procure an abortion, acts which he of course condemns in his theological work.[10] Among the fruit and vegetables that have contraceptive virtues, we will mention the pear, but especially the roots of the pear-tree, a tree in which, incidentally, the authors of medieval novels, in Latin or the romance languages, liked to set amorous frolics, especially if they were illicit.[11] This was doubtless no mere coincidence.

Medical discussions of contraception are fairly closely linked to discussions of diet: this or that foodstuff, by virtue of its qualities, has the property of encouraging or of preventing the production of semen. Thus all the foodstuffs that are hot and moist will favour the sexual capacities of the male. The logic of the way the world is represented, arranged according to qualities, means that each substance is given a definite purpose. Therefore sexuality is very commonly mentioned in diets, but without any great emphasis being laid on the sexual function, which is merely one of several functions. Every food product has its place in the tightly controlled, well-ordered system of the Galenic qualities, and the theoretical framework is even more restrictive than a modern dietetics – meant to govern every moment of one's life – would be. Contraceptives are explained in the same way; a product in which cold is the dominant quality may not only extinguish lust in an individual, but overcome the heat necessary for the action of the male seed. Thus numerous contraceptives proposed must have worked very efficiently, by modifying the vaginal pH. Certain procedures follow this explanatory principle, others follow the system of analogical thought; several others strike us as results of the mere chance of empirical discoveries, and certain have to do, as we have said, with magical practices.

The richness of the discourse on contraception is not merely to be ascribed to an intellectual curiosity restricted to the circle of practitioners or of the great university teachers. The *Thesaurus Pauperum*, attributed to the doctor and philosopher Peter of Spain, who became pope under the name John XXI, is for more than one reason an exceptional work.[12] It is a medical compendium, whose reputation enabled it to be incorporated into several reworkings; one cannot tell whether it is a scholarly work or a work that lists the prescriptions of popular medicine, since in this text – as must have been the case in everyday life – bookish knowledge and popular lore influenced each other mutually. The place given to prescriptions relating to fertility and sexuality is considerable: out of 116 of the prescriptions, 34 are for aphrodisiacs, and 26 for contraceptives, whereas 56 relate to means of ensuring fertility. To this must be added the numerous prescriptions that, meant to bring about menstruation, could be read as prescriptions for abortion. This 'reading' seems all the more plausible in that it is suggested by Rhazes; in his *Liber ad Almansorem*, the Arab doctor closes the list of remedies supposed to expel the embryo with a reference to the chapter 'De Menstruorum Provocationae'.[13] One must therefore add to the information on the means of procuring an abortion that was in circulation in the Middle Ages the innumerable methods which, under the pretext that they acted as a stimulant on a sluggish menstrual flow, could be openly divulged. To return to the *Thesaurus Pauperum*, it is noticeable that certain aphrodisiacs aim not only at increasing the man's potency, but also at procuring a surplus pleasure for the woman. Taken from the Cyranides, and from Dioscorides, there appear several products which, left hanging or placed somewhere in the house, are meant to keep demons and evil spells at bay. The protection of fertility, of the male's potency, is not, in the type of society we are concerned with, limited to the reproductive organs, or to physiological questions, but extends to a whole set of things and to the territory over which the male exercises his power and his rights of possession. Proof that this territory is vulnerable takes from the man all or part of his potency, of his sexual capacity. Behind magical practices is concealed a truth well known to psychology. Under the name *Trotula* there appears a bizarre prescription about which the author of the *Thesaurus Pauperum* expresses the greatest reserve: 'When the woman does not want to conceive again, she must place in the afterbirth as many castor seeds . . . or grains of barley as she desires to have years of sterility.'[14] It seems that the author is here referring to, without understanding, the use of the loop, put into position when a woman decides that she has had enough children. It goes without saying that the procedure must have been particularly efficient – indeed, too efficient, for the risks of infection appear to have been far from negligible. Among the means of facilitating conception may be mentioned a certain way of raising the legs which, repeated by Rhazes, shows that variations in positions existed for the noblest reason.[15]

We will be returning to the question of positions, but it has to be recognized that medicine was a powerful supporter of theology: for Avicenna had asserted that when the woman is on top of the man, the effort of expelling the seed can cause lesions in the male organs.[16] Almost all the medical indications that we have come across on the retention of semen and spatial localization in the womb were reasons in favour of the so-called natural position, and this position alone. It is not out of the question, as we will see, for doctors to advise certain variations, but for those same reasons. After intercourse, every muscular contraction, and every movement, is capable of making the liquid formed by the reunion of the two seeds run out. That is why sneezing and jumping around – the number of jumps being precisely defined – had been considered, ever since Hippocrates, to be a means of avoiding pregnancy.[17]

We will conclude by underlining this fact: in 1256 and throughout this period in general no great theologian, neither Albert nor Thomas, nor Alexander of Hales nor Richard of Middleton, accepted the decision of the *Decretals* to consider contraception as homicide. The severity of certain writers was not the rule. As far as abortion was concerned, the limit of forty days was established: if the embryo did not reach this age, the only penalty laid down was exile. There was room for a whole range of activities, which were not so severely repressed as one might have expected. A judicial sentence mentioned by Ernest Wickersheimer demonstrates this:

Adelheit von Stutgarten, an undesirable German woman nicknamed *the limper* (*die hinckende Artzatin*), had, in Sélestat, procured abortions for many respectable women by making them take concoctions or roots. On 8 August 1409, she was requested to cross back over the Rhine and had to promise not to return for three years unless she obtained special authorization.[18]

It was an extremely lenient punishment and the twentieth century has sometimes been heavier handed.

Theologians hounded sexual behaviour that did not coincide strictly with the act of reproduction in the context of marriage. As far as the sexuality of adolescents was concerned, the net cast was far more widely meshed. Medicine continued to proclaim the existence of physiological necessities and the right to pleasure. Noonan thinks that very little information about the sin of contraception came via the confessor. A whole strategy of discourse must have been elaborated, and priests were well aware of the information that they could involuntarily reveal. John Nider in his *Manual for Confessors* counsels caution when questioning somebody about sins against nature, 'lest something be disclosed to the simple of which they were ignorant', but 'With caution, the interrogation is to be made.'[19] We also have Albert the Great justifying the time he spends discussing the various positions, in his commentary on the *Sentences*, by the fact that he is driven to do so 'by the monstrous things heard these days in confession'.[20] The study of the documents relating to the way sexuality was regulated hardly enables us to uncover any more details about the sexual life of men and women in the Middle Ages. There was a gap between the desire to enquire – which was bound at the very least to arouse curiosity – and an altogether up-to-date body

of medical knowledge which clearly demanded more liberty. From the
thirteenth century, an art of love continued to develop: pleasure in itself
was the subject of a train of thought quite independent of reproduction.
The impetus was given by a philosophy based on the idea of nature,
but it seems that over and above the 'naturalists', a way of enjoying sex
over which official theology had little power continued to develop.

DARING IDEAS IN SECULAR THOUGHT

Throughout its history, the Church recognized in the sexual act only
the function of reproduction and a great many of its combats were in
defence of this basic belief. A number of heresies transmitted a turbulent
impulse that risked making the whole species self-destruct – that is, of
course, if we take the biological point of view. Courtly ideology, which
was a source of inspiration for French medieval literature and gave it
its subtlety of psychological analysis, constituted one way of keeping the
demands of reproduction at a distance.

The erotic of the troubadours

Because courtly love was extramarital, it was an accomplishment based
on the desire to control the impatience of instinct, on the successful
passing of a series of initiatory tests and finally on the discovery, thanks
to the *dame*, the lady, of a world of spiritual values. According to
troubadour poetics, the lady focused on herself her lover's emotions, the
activity of his imagination and the way he saw his world. In this,
satisfaction was proscribed, for the essential thing for anyone devoting
himself to the service of love was the narcissistic cult of his desire; the
woman who received this homage derived pleasure from the exercising
of her power and could only refuse the embrace that would have put an
end to it. We hope we will be forgiven for the brevity of this extremely
incomplete sketch of courtly love, envisaged as a functional ideal. From
these heights we have to come back down to reality as it was in fact
experienced. On the level of sexuality,[21] courtly love could only be an
art of cultivating arousal, and of prolonging its duration so as finally to
bring it to a conclusion. It differed from the erotic arts whose object was
the consummation of love in the best possible conditions, and which
consisted of a codified practice of coitus which allowed one to gain
knowledge of the cosmos and to discover the harmony between microcosm
and macrocosm.

In his quest, the courtly lover passed through a series of stages that brought him progressively into the intimacy of the lady. He was *fenhedor* (suitor), *precador* (suppliant), *entendedor* (recognized lover), and *drut* (carnal lover). Poetry could not and must not mention the consummation of love. Bearing this in mind, one may wonder what the texts say or allow one to surmise. There existed a whole series of artifices whose aim was to bring desire to a climax, among them the practice, so clearly attested in French medieval literature, of the *concubitus sine actu*. René Nelli has evoked the crucial moment of the quest for perfection, namely the encounter of the lover and his lady in the course of the final ordeal constituted by the *asag* or *asais* (= test).[22] This moment was both a recompense and a new temptation, and yet the satisfactions offered were merely rhetorical, as appears from the words of the Countess of Die: '[The lover] will thus have to be content with reclining on the cushion of his bare arms, while lying next to her (*jazer*); with gazing at her (*remirar*), with *tener, abrassar, baizar, and manejar* her (holding, embracing, kissing and caressing her): these are the substantial intimacies permitted by courtly love.' It would seem, according to the teaching of the *trobairitz* (women troubadours), that the lover never enjoyed the rights of the husband. Nothing that might recall the husband's prerogatives, the brutality of the procreative act, could be accepted in this exercise.[23] If the behaviour alluded to was indeed as described, courtly love seems to have led to a rediscovery of and demand for clitoral stimulation, to the use of the man as an erotic stimulant – and thus to a sexuality that was active, specific and, what is more, recognized and even exalted as such by an aristocracy of the sensibility and the intelligence. Taking its cue perhaps from Arab erotic lore,[24] although this is a controversial question, intellectual life was for the first time associated with women's demands.

History has recorded several philosophical endeavours based on controlling the emission of semen. In ancient China it was thought that this should happen as little as possible:

Thus in Chinese literature on sex the following two basic facts are stressed again and again. First, a man's semen is his most precious possession, the source not only of his health but of his very life; every emission of semen will diminish this vital force, unless compensated by the acquiring of an equivalent amount of *yin* essence from the woman. Second, the man should give the woman complete satisfaction every time he cohabitates with her, but he should allow himself to reach orgasm only on certain specified occasions.[25]

Retention was thought of as a kind of therapeutic gymnastics, but independently of this medical-religious value, polygamy made it necessary. The same social conditions existed among the Arabs and men paid the same attention to controlling their sexual activity. Whether it drew on outside influences or whether it was a spontaneous blossoming in a privileged milieu that had access to a certain type of knowledge, the form of erotic art adopted by the West was highly original: it was woman who was in control of pleasure. Though courtly love had a carnal component that could hardly be denied – a component that fitted in with the life of society at large – it remains nonetheless true that, repeated again and again in literature, not experienced but dreamt about, courtly love was ultimately able to become the manifestation of a certain fear of women, a slippery slope that led to a gynophobia of which the fabliaux are the most brutal, but perhaps the least harmful expression.

Let us return now to more down-to-earth considerations. Belief in woman's continuous fecundity, as well as the interest taken in lineage, made the emission of the man's sperm into the vagina more or less impossible. It would seem that in conformity with the unanimously accepted scientific theories, 'extra vas' emission had to be practised. It could take two forms: either a satisfaction achieved manually or with another substitute, or else, after the lover's probationary period, the practice of the 'amplexus reservatus', that is, of extravaginal ejaculation once satisfaction had been given to the woman. It is certain that if the aristocracy used this procedure, there must have been a great freedom of behaviour. It was a well-known contraceptive method since one finds it mentioned by Arab doctors. On the other hand, Noonan declares that 'Up till about 1480, amplexus reservatus had been ignored by many authorities, championed by a few, and attacked by none.'[26] In Noonan's opinion, the absence of vigorous discussion about it bears witness to ignorance of the practice. It is evidently the most elementary and least dangerous means of contraception, but one which presupposes information about, and communication with, the world of women. We are reduced to examining what was not said.

The enigma of the De Amore of Andreas Capellanus

Recent studies, to which we will have occasion to return, have tried to shed new light on the work of Andreas Capellanus. Many terms and allusions in it are said to include an erotic meaning. Before undertaking an exploration of the Latin text, one should appreciate the spirit of this work, with the help of a translation that was carried out a century later (1290) and which exists only in the form of a single manuscript. The author of this translation, Drouart la Vache, treats not only the literal meaning of the text, but also the content and composition of the work, rather casually: additions and omissions abound. On hearing the Latin text, the reaction of Drouart la Vache was, to say the least, curious: 'When I had seen the book and my companion had read me a little from it, you must know that the subject certainly pleased me a great deal – indeed, it pleased me so much that I began to laugh.'[27] The translator's hilarity is somewhat surprising, for this treatise is considered to be an epitome of courtly casuistry. It lists the types of behaviour that are acceptable, when there is love between two people from different levels of society, and it contains the famous judgements of love that concern difficult cases: fictitiously or not, noble ladies give their opinion on the conduct to be followed in conformity with the rules of courtly love. The translator, furthermore, took some persuading before he would set to work, and he feels the need to take oratorical precautions, declaring that, if he happens to say gross things – which his book is concerned with – it is not his fault, but his subject's.[28] The end is even more surprising, for it reveals the love of playing games which extends from beginning to end of the text. The translator tells us his name in the form of a riddle. As for the passage just before this, he is suitably mysterious, but it cannot be denied that he is inviting us to seek out information hidden in the text itself:

And I feel quite at ease in telling you all this, for I have written this book, which is of good origin and good tone, as it should be, for clerics and not for the laity, who are rather simple and ignorant, for in the book there are several ideas which the laity could not understand, even if one threatened to drown them or hang them.[29]

Clerics who think about it will understand the text perfectly well and will gain pleasure from it. It is clearly difficult to fail to recognize the meaning of warnings such as these, and unjust to criticize anyone who cares to make an attempt to find out more about the real meaning that Drouart la Vache wanted to give the text he translated.

One of the work's constant features is thus an encomium of initiates, that is, of the clerics; their way of loving puts them above all classes of society and they have much to teach others. From the arguments that seem to establish the superiority of the cleric, in the traditional debate that sets him against the knight, we will choose an exchange, inserted by Drouart la Vache, between the master and his disciple. The disciple talks about the love that can exist between nuns and clerics and reproaches the master for speaking ill of these nuns, when he has just confessed that he has been more than kindly disposed towards one of them. The master replies with a long didactic discussion, the terms of which should be noted:

Car en cest monde puet avoir	For in this world there may exist
Double amour, ce dois tu savoir	A double form of love, and this you must know.
La premiere est pure apelee	The first is called 'pure love'
Et la seconde amour mellee.	And the second 'mixed love'.
Cil qui s'entraiment d'amour pure	Those who share 'pure love'
Dou delit de la char n'ont cure	Pay no heed to the work of the flesh
Ains wellent sanz plus acoler	But want merely to embrace
Et baisier sanz outre couler.	And to kiss each other without going any further.
Et tele amour est vertueuse,	And such a form of love is virtuous
Ne n'est a son proime greveuse.	And is not harmful to one's neighbour.
De tele amour vient grant proece	From such a form of love springs great prowess
Et Diex gaires ne s'en courece.	And God is hardly angered at it.
Et tele amour puet maintenir,	And such a form of love can be practised,
Sans li por grevee tenir,	Without the woman feeling afflicted,
Pucele et fame mariee,	By virgin and married woman,
Et nonnain a Dieu dediee	And nun devoted to God.[30]

We have deliberately given a literal translation. There is another way of understanding the line 'Et baisier sanz outre couler': one can, of course, read 'Make love without ejaculating' or more exactly 'without shedding anything more than the secretion of the prostatic humour'. In the line 'Ne n'est a son proime greveuse', the adjective *greveuse* has been interpreted as meaning 'harmful', but it must have the meaning 'capable of causing pregnancy'. 'Being pregnant' is a well-known meaning of the past participle. It can now easily be understood why virgins, married women and nuns can indulge in this form of love without considering themselves to be 'grevée' – harmed, or made pregnant. The meaning of this passage appears to us to be most explicit.

A discussion which pits 'the most noble man' against the 'noble woman' brings forward further arguments. The debate has in fact to do with jealousy, and the man declares, in perfect conformity with courtly tradition, that love does not exist between married people. One must clearly choose between love 'ou l'em puist faire sans peür sa volentet tout asseür' ('love in which one can carry out one's desire without fear, in all security') and the other form, which we have alluded to. The text declares without beating about the bush that if, within marriage, it so happens that a man feels *affection* other than that which is necessary to have descendants – or if his wife asks him to behave in this way and he obeys – it is a grave sin, it is adultery. A strange piece of casuistry! This codified behaviour on the part of the married man clearly enables one to suppose that there is behaviour proper to courtly love, and in the light of what is said in the text, we will attempt to interpret a long comparison that comes before it:

Se l'yave de l'eve bien clere,	If the water of a clear stream
Par. I. chanel commence a courre,	Starts to flow along a channel
Qui est plains de boe ou de pourre,	Which is full of mud and dust,
L'yave, qui de sa nature	The water, which of its nature
Estoit clere, devient oscure	Was clear, becomes dark
Por la gravele qui li donne	Because of the gravel which gives it
Oscurté . . .[31]	Darkness . . .

The husband who behaves as a lover perverts the courtly intention (the clear water). That is the literal meaning of this allegory. One may also wonder whether the clear water is not the prostatic humour, which becomes a spermatic discharge because of the semen, *la gravele*, a word that must be linked to the group *grever* that we have just mentioned; as for the Latin text, it has the word *arena*, i.e., *a-rena*, 'coming from the kidneys' (?). The reader will have to forgive us for these guesses, but we do not think we are betraying the spirit of the clerics in indulging in them, especially when they themselves provide us with a number of other examples.

The boldness of the interpretations that we are suggesting may seem surprising, for we are basing our interpretations on fragments of texts examined brutally under a garish light. In fact, they are put in perspective by a context of ambiguous expressions which it would take too long and be too tedious to go into here. The whole work is devoted to the problems of sexuality. Thus one theme of the discussion is returned to again and again: should a woman prefer an adolescent to a man of mature years?[32] The answer is this: the lady has to prefer the older man, for it may so happen that the young man never retains what the lady teaches him, so that he will never be *wise* (*sage* – suitable for the courtly game of love?). Indeed, says the text, if you have any light of understanding, you know that seed cannot always bear fruit, just because it has been scattered onto the ground (in Latin: 'quia non semper jactata producunt semina fructum').[33] We are clearly dealing with the image from the Gospels, repeated a thousand times, of the seed cast either onto good ground, or onto stony ground, but, in this context, it cannot fail but remind us of the act of Onan, the father of *coitus interruptus*, who 'shed his seed on the ground', as the text of Saint Jerome put it.[34]

Certain questions are referred to bluntly in the text, such as the insatiability of some women. The lines can be quoted without further commentary.

N'aies dont de tel fame cure	Do not bother over such a woman,
Se tu ne pues tant tribouler	If you cannot work away
Que tu la puisses saouler.[35]	Enough to make her satisfied.

There are other questions, equally realistic, for instance that of the man deprived of the 'instruments of nature'.[36] Likewise, the rights of the initiating woman are mentioned: the man theoretically belongs to the woman who has awoken him to amorous life, or rather to sexual life, according to courtly tradition, since success comes with *baisiers et par acolees*, 'kisses and embraces'.[37] The way the cleric behaves with a woman from a lowly family is one of the most brutal things ever said: 'And if it happens that you are seized by the desire to love a woman from a lowly family, and if you can find a propitious occasion, you must not restrain yourself, but take your pleasure without waiting for another opportunity.'[38] This is tantamount to advising rape. The contempt with which villeins are treated is horrifying: they carry out 'the act of lust like animals'; in other words they limit themselves to satisfying their instincts. One must abstain from teaching them the art of loving – *the doctrine* – since the economic results would be disastrous: the fields and vineyards would lie fallow. It is clearly difficult to imagine that the secret in question is the art of refined courting – it would seem rather to be the art of controlling the sexual instinct, which the author always calls 'the work of lust' – in opposition to 'doing one's will', which presupposes a subject who is always in control of events. Courtly love is aristocratic – that much is clear. In its excessiveness, it disregards the frontier between the human race and animals, setting up a new boundary between courtly lovers and the rest, who are considered to be no different from animals. This is yet another reason for condemning it from a theological point of view. This superiority, established on the basis of the way one carries out the sexual act, shows on the part of the clerics a completely unjustifiable contempt for 'villeins and carters'. The cleric,

FIGURE 3.1   *A detail from the Bayeux tapestry, showing a rape (?) scene. (Reproduced with special permission from la Ville de Bayeux)*

the intellectual, is heir to the ideology of the civilization of the *pays d'Oc* and makes use of it to set himself at the top of the social hierarchy – and if our reading of the text is accepted, this feeling of superiority, proclaimed with such arrogance, finds its justification in the control of the sexual instinct.

The practice of *coitus interruptus*, as we have revealed in certain passages of the text, enables us to resolve two perplexing difficulties concerning the exact nature of courtly love. Hardly anyone still believes in the chastity of courtly relations, and most historians of manners and of literature agree that the woman, after a longer or shorter probationary period, granted the *guerredon*, the ultimate favour – that is, the gift of her person. Thus Moshe Lazar, drawing on the text of Andreas Capellanus, declares: 'There are four stages in the lover's conquest and they depend on the lady: 1) inspiring hope; 2) promising a kiss; 3) the embraces; 4) the surrender of her whole person'.[39] He also contrasts the

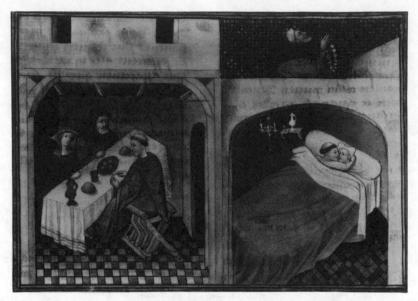

FIGURE 3.2   *A monk and a woman in bed. From the* Decameron. *Ms Arsenal. 5070, fol. 108v. (Bibliothèque Nationale, Paris)*

two forms of love: 'In reality, *purus amor* and *mixtus amor* are two forms of the same love, two ways of expressing the same passion.' Let us recall the definition of pure love, a love that does not end in the performing of the *final act of Venus*. This must be understood as ejaculation, and it is here that the line between the two forms of love has to be drawn. Such a practice enables satisfaction to be obtained without risking pregnancy. Medieval medical texts taken as a whole show that the woman is continuously fertile, except when she is pregnant or suckling. It is difficult to imagine the courtly lover making the most of these happy times. Every performance of the sexual act involved the possibility of impregnation and one thus had to envisage the possibility that the line of descendants might be broken, and that there might thus be one or more bastards among the children. It does not seem that this was accepted by medieval society. The practice we have just alluded to allows us to resolve this contradiction.

If we agree that Drouart la Vache in his translation is referring to *coitus interruptus*, a second contradiction also has to be resolved. This was realized very clearly by Noonan. The Cathars are alleged to have practised this form of contraception, as did the aristocracy of the *pays d'Oc*, as is confirmed by R. Nelli. Later, this practice is supposed to have fallen into disuse, as Noonan says: 'This method of avoiding procreation, untreated in the thirteenth century, is launched under fresh auspices, and as a solution to a different problem, in the fourteenth century.'[40] To smooth over this break, the author supposed that knowledge of this procedure had accompanied the Cathars who settled in Northern Italy and that, from this region, it spread out in a second wave. The absence of texts dealing with the subject means nothing, and it is completely impossible to believe that a practice concerning an important part of society could have fallen into oblivion. Furthermore, aristocrats in the north were acquainted with it well before the Albigensian crusade, since it was a component of courtly love.

Nobody will deny that Drouart was well informed. He refers to the '*naturiens*' or 'naturalists',[41] that is, to the very open intellectual milieu of the Faculty of Arts, steeped in Aristotle and particularly inclined to discuss problems of sexuality. It is certain that the Aristotelian concept of the respective roles played by the male and female seed, by giving an outright supremacy to the male sperm, led, in extremely liberal circles, to new trains of thought on the subject of contraception. This now needs to be verified: if it can be agreed that Drouart was talking about *coitus interruptus* in his translation of the Latin text, is any reference to this practice made in the original?

We will be granted as a working hypothesis that the characters of different social conditions may represent the different degrees of information concerning erotic technique. Here are the terms in which the dialogue between a nobleman offering his services to a noblewoman are couched: a dialogue, that is to say, 'between initiates'. We quote the translation, though the Latin text is even more expressive:

The woman says: It is very easy to discover the entrance to the court of Love, yet difficult to abide there, because of the pains poised above lovers; but people find it impossible and insuperably hard to leave because of the acts of love which they crave. Once he has truly entered Love's court, a lover can say yea or nay to nothing except the fare placed before him on Love's table, and that which can please his partner in love.

So a court of that kind we should not approach; we should totally avoid entering a place from which it is not possible to leave freely. Such a place is comparable to the court of Hell, for though the gate of Hell stays open for anyone desirous of entering, once inside there is no prospect of leaving.[42]

The erotic reading of such a text as this cannot be avoided. Furthermore, the allegory that follows adds new and precise details:

The story goes – and it is true – that at the centre of the world is set a palace with four highly ornate sides, in each of which is a most beautiful entrance. . . . The eastern entrance has been reserved for the use of the god of love alone, whereas the other three are specified for particular classes of ladies. The ladies of the south entrance always linger at the open doors and are for ever to be found on the threshold. Likewise the ladies of the western entrance, only they are always to be found wandering outside the threshold of the door. But the ones who have the privilege of guarding the north entrance always remain behind closed doors, and observe nothing outside the boundaries of the palace.

Lest we lose ourselves in architectural reveries, the author carefully unveils the meaning of each of the gates:

The ones who for ever linger at the open gate, and are always found on the threshold, are the women and ladies who, when a man requests entry, are skilled at making careful enquiry of the meritorious treatment deserved and the honest character maintained by him who seeks to enter the open gate. When they are fully confident of their merits, they admit those entitled with every honour, but then repulse the unworthy far from the court of Love. The ones who claim a place at the western entrance are the promiscuous women who refuse no one; they admit all indiscriminately and are available for all men's pleasure. But those assigned to guarding the north entrance, who remain perpetually behind closed doors, are the women who open to no man's knock and refuse all men entry to the palace of Love.

A distinction is established between three types of women: courtesans, chaste women and courtly lovers, who do not refuse entry into the palace of Love, on condition that the lover first gives proof of his mastery. The dialogue which crowns book I of the De Amore brings together a fully initiated person and a lady who is in no way inferior to him. Nobody will be surprised that at the summit of the hierarchy of lovers is placed the cleric; the argument put forward does not contradict our interpretation of the text:

The cleric is seen to be more careful and wise in all things than the layman. He orders himself and his affairs with greater control, and is accustomed to governing everything with more fitting measure; because he is a cleric he has knowledge of all things, since scripture gives him this expertise. So his love is to be accounted better than the layman's, because nothing on earth is established to be so vital as that the lover should have experience in diligent application to all things.[43]

Everyone has happily recognized in the De Amore of Andreas Capellanus the expression of a technique of love.[44] The practice of coitus interruptus, with the physical and moral control that it presupposes, does seem to be the main lesson taught by the work, from the twelfth century onwards. The doctrine was received by his translator Drouart, although he did not express this half-secret lore in the same way. A recognition of these facts does not mean that those interpretations which see in courtly love a spiritual quest have to be excluded, but it does modify them somewhat. It is now easier to see the kinship between courtly ideology and the great erotic treatises of other civilizations, and demonstrate that, although they used different techniques, there was a profound analogy between them in their philosophical and religious quest. This is the deep meaning of a whole section of the poetry and courtly literature of the *langue d'Oc*. On the other hand, one can also emphasize the way the clerics appropriated these techniques. Inspiration gave way to rhetoric, the philosophical quest to the assertion of a power exercised in the very heart of society. Furthermore, a certain way of handling language and playing on words, completely infiltrated the way these intellectuals discussed love. Everything in their language seems to conceal a trap and over and above the explicit meanings that we have just elucidated were expressed, in ambiguous words, the erotic fantasies of initiated scholars – great masters of quibbles and rhetorical figures – who cheerfully transgressed the boundaries of what was permitted.

For a medieval scholar any commentary, any exegesis, began with a literal understanding, with a word-for-word reading; according to the Isidorean tradition, the word is in itself the bearer of a hidden truth which it often reveals only at the price of a modification or fragmentation of the signs of which it is composed. This level of reading cannot be neglected in Andreas Capellanus. Recently, Betsy Bowden has shown the importance that had to be attached to the sound of words and the parallels that could be drawn between different syllables.[45] Following a tradition firmly rooted in lyric poetry, in genres close to the *Carmina Burana* or in the Latin comedy of the twelfth century, puns are used so as to suggest, implicitly, a completely different meaning from that which is made explicit. This is revealed particularly in poetry of homosexual inspiration which, furthermore, has recourse to a vocabulary comprehensible only to initiates.[46] In his *De Planctu Naturae*, Alan of Lille deplores the rupture of the *copula verborum*, the confusion of genres that link moral perversion and the corruption of the arts of expression.[47] The influence of Cupid perverts poetic meaning; the God of Love uses antiphrasis to make words lose their normal associations.

Betsy Bowden is merely pointing out that which is most visible in the *De Amore*: the association of certain syllables, the astonishing frequency and the repeated grouping of words that could refer phonetically to a sexual meaning, playing also on the vocabulary of the vernacular. The American scholar thus picks out a series of terms, such as *poenis, cunctus, penitus*, etc. Hubert Silvestre adds to this list *probitas* which recurs 130 times in the treatise and which Drouart la Vache translates as 'prowess'.[48] This quality that the woman must seek in a lover is the opposite of 'biauté', *forma*, a specifically feminine attribute, as the poet Serlo of Wilton reminds us: 'Cur probitas maribus, cur virginibus data forma?'[49]

Apart from these puns which, as one has to recognize, do not differ from the tradition of medieval jokes,[50] Betsy Bowden notes the use made of metaphors already present in the erotic vocabulary of classic Latin: *via, porta, mors, semita, tractare, arma, locus,* etc.

The duplicity of the work is, furthermore, revealed at the end, when the author addresses the recipient: 'So if you peruse with careful diligence this teaching of mine which I send you within the covers of this little book, it will present to you two differing views *(duplicis sententia)*.' The reason clearly expressed is that while the first part teaches the way in which to 'obtain in full measure all bodily pleasures', the second constitutes a strict warning. It is the discussion of this reason that is referred to notably by those who hold to the idea that there is a 'pre-Averroist' flavour to the text.[51] As Betsy Bowden has shown, this cannot seriously be maintained, because the second part contains as many puns and sexual allusions as the first. The work taken as a whole presents – admittedly in such a way as hardly to be noticed – formulae designed to awaken the reader's attention; for example the phrase: 'By lending a diligent and attentive ear to the advice which I have composed . . . you can become fully acquainted with the art of love'[52] invites one to pay special attention to phonetics.

If one has to look for a hidden meaning, puns are not enough; they are often meant merely to attract attention and to provoke the mirth of a Drouart la Vache. There are other allusions that one perceives if one reads the text 'with careful diligence', as Andreas Capellanus puts it – allusions that appeal to the culture of the medieval cleric. For the *De Amore* presents us with a strange bestiary, composed for the most part of winged creatures. We will take as an example the partridge, mentioned on several occasions. Before beginning our reading, we must refer to the 'dictionary' of the Middle Ages, that is, the *Etymologiae* of Isidore of Seville; here is what is said about the partridge:

Partridge gets its name *de voce*. It is a filthy rogue of a bird. For male mounts male and blind desire forgets gender. [This bird] is such a rogue that it steals another bird's eggs to hatch them, but this piece of roguery bears no fruit. When the young hear the voice of their own mother, they abandon, out of a certain natural instinct, the bird that has hatched them, and return to their mother who bred them.[53]

The *Bestiaire d'Amours* of Richard de Fournival described at great length how infertile partridges managed to hatch eggs. Bartholomew the

Englishman took from Isidore the notion that they indulged in homosexual relations. The two authors that we have just quoted [54] suffice to prove that the description of the way of life of this bird did not remain a dead letter in the Middle Ages. Furthermore, if one refers to the correspondences in erotic classical vocabulary, it will be found that the partridge is associated with *paedicatio*, that is with anal practices.[55] Let us now reread one of the passages in which the partridge appears in the *De Amore*. The allusion is found in the course of a dialogue between the commoner and the most noble lady; to be admitted into the court of love, the lover from the lower classes must more than any other give proof of his merits

because an aspiration avowedly beyond an individual's nature is usually dispelled by a puff of wind and lasts only a moment. We are told that occasionally among kestrels are born certain birds which by their courage or fierceness subdue partridges, but because this achievement is acknowledged to be beyond their nature, people say that such fierceness cannot endure for more than a year, reckoning from their birth. So a commoner can be selected for love by a woman of the higher nobility only if after considerable trial he is established as worthy.[56]

In this passage, which speaks of 'an aspiration ... beyond an individual's nature', the Latin text multiplies the possibilities of a double reading. Thus *ultra naturam* may mean 'against nature', with all that that implies when sexuality is being discussed, but also, giving *natura* a meaning that is attested in Cicero and which we have met with in the *Pantegni* of Constantine the African, 'beyond, on the other side of the nature [of the woman]'. The single expression *ultra naturam* includes two types of allusions that lead one to the same meaning. Let us add that the ferocity of the kestrels cannot last for longer than a year: *usque ad annum*. To designate what Walsh translates as 'kestrels', Andreas Capellanus uses the expression *lacertiva avis* in which the first term refers to *lacertus* (muscle) and the second requires no explanation, if one remembers how widespread its erotic meaning is in different languages, including those of the Middle Ages. Let us now return to the partridge: apart from the way of life described by Isidore of Seville, we find that Andreas Capellanus associates this bird in other allusions with the pheasant, *fasianus*.

The capturing of partridges by the 'kestrels' cannot last; it corresponds to a 'probationary period'. If it is still possible to doubt that Andreas Capellanus is suggesting in this way a channel without any risk, the rest of the same dialogue gives us other indications: the lady reproaches her lover with appearing to 'confound the order and the course of nature'. Several lines further on, it is announced that 'Abusive tongues cannot remain within courtesy's threshold'; the Latin text is more specific: 'Maledici intra curialitatis non possunt limina permanere', that is to say,

'Slanderers cannot remain within the gates of courtliness.' Here is what Drouart la Vache tells us about slanderers:

Por nule parole qu'elle oye	Let her listen to no words,
Ne por chose que mesdisant	Nor to anything the slanderers
Voisent par derriere disant;	Will say behind;
Car li mesdisant ont maniere	For the slanderers have a way
De parler touz jours par derriere	Of always speaking behind
Por faire a bons	To prevent the good from their
empeschement . . .	doing . . .[57]

The example of the partridge enables one to establish a range of probable meanings, which are backed up by other signs as well as mere puns; numerous other pages multiply the possibilities of double meanings so as to suggest what the 'freedom to love' is, according to Andreas Capellanus: knowing how to choose, 'at the crossing of two paths', the one where one must 'serve Love',[58] according to one's degree of knowledge.

This reading of the *De Amore* – which does not, of course, exclude other interpretations – would be of only trifling importance, if the work had not gained the success that is known and if it did not figure among the books condemned in 1277. The treatise of Andreas Capellanus cannot be considered as an accident, as the result of the fantasies of a single individual. It thus transcends the simple level of a joke. Scientific information surfaces in certain places: the name of the doctor Johannicius is quoted and the definition of love is hardly any different from the definition of coitus given by Constantine the African. One reads that going with courtesans does not require 'doctrine' (*doctrina*), or 'knowledge' (*sapientia*); indeed, as William of Conches teaches, such women are sterile. As a good cleric, Andreas Capellanus shows perfect control of the art of rhetoric; it can adorn his rhetorical figures in the colours of Venus, that is, subject them to 'inversions' or 'discolorations', which 'lead them to vice', to use the terms of Alan of Lille.[59] The author makes use again and again of *annominatio*, which consists of reproducing the same word, changing only one or two letters, or in setting next to each other terms that are identical in form, but different in meaning. The recourse to *translatio* that we have mentioned with regard to Isidore of Seville is equally evident; Geoffrey of Vinsauf, writing a little later than Andreas Capellanus, gives the following definition of this 'colour': 'There is *translatio* when an expression is transferred through similarity of its proper meaning into an improper meaning.'[60]

Treatises of poetics and rhetoric flourished at the time of Andreas Capellanus. Special attention must be given to the *Ars Versificatoria* of Matthew of Vendôme, also the author of a long poem which, under the title *Milo*, is extremely scabrous in content.[61] An examination of the *Ars Versificatoria* will come up with some astonishing examples. Edmond Faral had himself pointed out that the statement of the rules meant to govern the physical description of the characters constituted highly favoured terrain for rhetoric to lend 'the veil of its figures to licentious *pointes*'. Let us look at the paragraph illustrating the description of the facts and the three attributes that are thought to explain the circumstances that precede, accompany and follow the event described by the poet:

One can give a familiar example of these three [attributes]:

> 'Risus amor, coitus, ventris conceptio, triplex
> Indicium laesae virginitatis habent.'

There follows an analysis of these lines (which mean: 'Love welcomed with laughter, followed by intercourse, and then by the womb's conceiving, constitute a threefold sign that virginity has been lost'): *risus amor* is called *ante rem*, for a *consensus* of the mind (*mentis*) is the prelude to pleasure; *coitus* clearly constitutes the attribute *cum re*, while *ventris conceptio* is the consequence, the attribute *post rem*. The insertion of the words *risus* and *mens* in this context may also lead the reader to a metaphorical interpretation. The advice given by Matthew of Vendôme on the subject of describing the attributes of the feminine sex are even more explicit; apart from the fact that the word *forma* (and not *formositas* or *pulchritudo*) appears in a position of some importance, the portrait of the matronly woman is followed by an unexpected definition of *libido*: 'Est autem libido res vilis et turpis ex vili et turpi membrorum agitatione proveniens, cuius appetitus plenus est anxietatis, satietas p[l]ena paenitatis' – 'Desire is a foul and filthy thing arising from the foul and filthy stirring of the members: its craving is full of anxiety, and once satisfied it inspires only feelings of guilt'. This puts us in a specifically carnal context and at the same time gives us useful parallels for the words *anxietas* and *paenitas*.

The definitions relating to grammar or rhetoric frequently provide Andreas Capellanus with an opportunity to give clerics, who alone are capable of understanding his allusions, the key to a double reading. The grossest puns are merely signs aimed at alerting the reader's attention to the erotic content of the work. To judge the complexity of the system used by clerics in the Middle Ages, we will content ourselves with referring to the passage devoted to *equivocatio* in the tradition of Drouart la Vache. We will conclude with one of the facile jokes with which he again pays homage to the cleric:

Car fames, por nules raisons,	For women cannot for any reason
Ne pueent le comperatif	Pass the comparative,
Passer, mais ou suppellatif	But to the superlative
Pueent mout bien monter li homme	Men may mount most easily,
Et estre tres noble, si comme	And be most noble, just like
Font clerc.[62]	The clerics.

The Roman de la Rose

We cannot discuss eroticism in the Middle Ages without mentioning the *Roman de la Rose*, or rather the continuation of the work of Guillaume de Lorris by Jean de Meung, the composition of which can be dated between 1275 and 1280. The work came into being right in the middle of the upheaval caused by the spread of Aristotelianism and Averroism. Siger of Brabant and Boethius of Dacia, representatives of this current of thought, had provoked in the Faculty of Arts in Paris an intellectual effervescence, to which Jean de Meung, like the good Arts man he was, was doubtless no stranger.

For the revolt was not only intellectual and dogma really was being questioned, as Fr. Mandonnet shows:

It was clearly in that turbulent and unbridled milieu criticized by the legate that one could hear remarks such as these: theology is founded on fables; the wise men of this world are the only philosophers; Christianity is an obstacle standing in the way of science; the only happiness is in this world; death is the end of everything; one must confess only in appearance; one must not pray; fornication is not a sin.[63]

Such a climate evidently encouraged a free and easy way of life, and the right to pleasure that the *De Amore* of Andreas Capellanus so insistently laid claim to – with the difference that there was now an intellectual authority who enabled one not to fear punishment any more, who diminished, and even eradicated, the feeling of guilt. Averroism was exploited not only by intellectuals, but also among the people; it encouraged a remarkable freedom: it appealed to the idea of the unity of the intellect, the unity of human intelligence. Let us again quote Fr. Mandonnet:

Of all the doctrines of peripateticism capable of wrecking the Christian faith, none was more disastrous in its consequences. With the denial of the personal survival of the soul after death, it suppressed a fundamental part of Christianity and as a consequence destroyed the rest. Thus neither the Christian philosophers such as Albertus Magnus and Thomas Aquinas, nor the ecclesiastical authorities, made any mistake about it.

Jean de Meung was not unaware of this new dimension of moral life and followed his own train of thought on love. Right from the start he claimed the right to freedom of expression, unless he was merely being facetious.[64]

We cannot possibly study in its entirety the erotic play in the *Roman de la Rose*.[65] It is a song of seduction in the course of which metaphorical procedures are deployed in all their richness. It is, furthermore, striking to discover that, from India to the France of the thirteenth century, poetry borrowed as much from the floral domain as from the bestiary, and drew on certain constant factors. The use of the myth of Pygmalion is revealing as far as the author's thoughts on the subject of women are concerned. The sculptor gives life to the statue, the lover 'warms up' and 'softens' the stone creature. As Daniel Poirion puts it: 'The lesson in sexual education here finds an eloquent illustration, the role of male initiative in the arousing of female sensuality being clearly indicated.'[66] Nonetheless, the conclusion is far from this delicacy and the Goliard-like vein keeps breaking through, reinforced by a copious supply of more or less brutal comparisons. Jean Gerson judged certain passages to be sacrilegious. There are even jokes in more than dubious taste. The essential thing lies in the ostentation of virility, which radically separates this text from the courtly current as defined in ideal terms. One might even sense a certain regression, as compared with the *De Amore* of Andreas Capellanus, who on several occasions foresaw the possibility of women taking the initiative, in particular when he advised that the woman should be able to 'speak' first. The influence of the Aristotelian current is doubtless not unconnected to this difference in language. The *Roman de la Rose* has given rise to a variety of reactions:[67] Martin le Franc was able to say that Jean de Meung saw in love-play merely the pleasure of 'tupping' and that he behaved, with the rosebud and the tender rosebush, 'like a drunkard full of barley beer'.

Presented in this way, the myth of Pygmalion seemingly responds to the desire shown by the clerics to affirm their pedagogic vocation in all things and to exercise their power through a kind of knowledge that they insistently claimed to hold a monopoly on. The other side of the story could be called the Pygmalion complex. It runs through all literature: the man must find the woman to whom he will reveal the pleasures of love, and whose sensuality will thereby be awakened. This complex haunts the brain of every man who dreams of the virgin over whom his power and sometimes his knowledge can be exercised. All is supposed to go well right up to the crucial moment: then comes a gap in the story and the *Roman de la Rose* is no exception to this rule.

Many systems see the sexual act as a way of knowledge, and integrate it into their whole scheme of thinking. Woman is declared to be a means of participating in the harmony of the macrocosm. In the *Roman de la Rose*, the sexual act is violence, sleep and death, and awakening. The last line of the poem speaks of coming round to consciousness again. If readers merely see this as the poetic telling of a story of deflowering, they feel cheated – unless they reflect that they have been able to cross a forest of symbols and that, once they have read to the last line, they have reached a higher state of awareness which will enable them to reread the text for its alchemical meaning, so as to penetrate to a dimension in which they will gain a new understanding of the cosmos. To reduce the text to a single meaning would condemn it to death. Its meanings are hidden and the means of gaining access to them have to be discovered. Since no philosophical or religious path to understanding is openly revealed, the veil must be lifted before the seeker can read of the union of the male and female principles: this constitutes another way of knowing and understanding the world.

The *Roman de la Rose* also includes a 'De Ornatu Mulierum', although it is sketched only very rapidly.[68] Advice is put into the mouth of the character of the old woman, and it is meant for the young lady lover. One can learn about the use of make-up and ointments, of artificial hair; there is an art of wearing a dress which is cut low both in front and at the nape of the neck, so that the whiteness of her skin may appear; a rather plump bosom must be supported with the help of a tightly wrapped piece of cloth. Ovid's influence can be strongly felt throughout the passage. There is something derisory and saddening about the advice given by the old woman: she has taken too long to acquire her precious experience and has not herself had time to put her knowledge to any use. She at least wants the young woman to make the most of this knowledge. But her lesson is dreadful in its realism; the woman must above all learn to 'pluck the pigeon', that is, 'fleece the simpleton'. In fact, from our point of view, what the text does not say is that it is the function of the old woman to be in possession of the art of love, whose ultimate secrets are never unveiled. The author sets up a strategy of erotic discourse, which functions in the literary fiction as an ideal to be followed, despite its rather caricatural character. In the world of women, there is clearly an exchange of information between youth and experience,[69] but afterwards this information is communicated to everybody, in the form of a work of literature. Exceptionally, and this is perhaps one of the consequences of the liberation of morals, by the end of the thirteenth century the conditions of discourse were revealed and the male reader could observe a game in which he did not take the initiative.

In his *Tree of Knowledge*, Raymond Lull creates the character of an old mother whose cynicism is quite disturbing. She is asked why she reproaches her son with his debauchery, whereas she tolerates it in her daughter: "'Lustful old woman," said the hermit, "why do you criticize your son for his debauchery, but not your daughter for hers?"'[70] Raymond Lull gives her this reply:

The old woman replied that her son was consuming his body and spending his money on [indulging his] lust, whereas her daughter earned money. Furthermore, her daughter indulged her lust before her very eyes, which her son never did. Finally, she could speak about lust without shame to her daughter, to whom she recounted the pleasures that she had experienced with men, pleasures that she was ashamed to tell her son about.

The old woman takes a certain pride in her own life of debauchery; it has enabled her to assert her liberty. But this brief dialogue also proves that the art of loving is transmitted by women and that men play no part in it, even within the family. The information that a boy learns from his elders will only create complexes: and this information will be riddled with errors if he learns it from companions of his own age. In the field of communication and of sexual initiation, the text reveals that a young man and a young woman are not in the same situation. Perhaps there is, in this extreme case presented to us by Raymond Lull, a revelation of the way information works in the midst of the family structure. The woman who speaks of her pleasure has in fact learnt all about the art of love, whereas the male has only ever been initiated into the function of reproduction. One can thus wonder whether the frustration of the male, his inability to understand – something that is, furthermore, translated by a title like the *Secrets of Women* – will not be a component part in the creation of the character of the witch. This developing belief focuses specifically on the older woman, who is accused of procuring abortions. If the person accused of black magic is often a woman, this is because there has been an increasing lack of communication between men and women, aggravated by male fantasies and by a feeling of guilt. Standing in as a substitute for direct communication between the two sexes, the alleged witch fulfils the deadly role of a go-between, and the only means that a male society possesses of expressing its terror, is to send her straight to the stake.

The features of the *vetula* (the procuress and occasionally witch) that we have just referred to, can be seen in that typical character of Hispano-Arabic literature, the *alcahueta*. The first occurrence of this word, of Arabic origin, in the Spanish language dates from 1196.[71] It designates the procuress, the madam. One of the most famous illustrations of the character was the *Celestina*[72] of the converted Jew Fernando de Rojas; of this play, published for the first time in 1499, Cervantes wrote 'that it would be divine, if it were not so human'. Celestina was at first a prostitute; she learned her art – which she calls a profession, that is, something based on a technique – from an older colleague. She boasts that after a year, she knew more about her job than the woman who had initiated her. Then she became a procuress: her house had room for nine girls between fourteen and eighteen years old. The list of her other jobs is particularly revealing: sewing, selling lace and ribbons, perfumery and a little pharmacy. She is seen selling creams and ointments and recommending fumigations, but she excels above all in the art of making women virgins again. Her activities as a procuress often begin with her delivering herself of a medical consultation. The image of Trotula, which appears explicitly in Chaucer, is clearly an underlying factor. Indeed, the treatise which the lady of Salerno wrote for women mixes gynaecological information with advice on beauty care, as we shall see. Long passages also describe ways of pretending one is still a virgin. If in the Middle Ages people never doubted that the author of the treatise ascribed to Trotula was a woman, this is because they came across characters who mastered a similar range of knowledge in everyday reality. To come back to Celestina, it has to be admitted that she probably did not know how to read; she merely passed on what older female colleagues had taught her. This suggests another source of information: the procuress boasts that she has all the ecclesiastics, from the bishop to the sacristan, eating out of her hand. Clerics have access to books; they know the 'secrets' and can divulge them to the women capable of putting them into practice or of teaching them.

The mother described by Raymond Lull says that her 'daughter indulged her lust before her very eyes'; likewise for Celestina, it seems normal for a procuress to be present at lovers' frolics. Outside the framework of prostitution, there was another character who witnessed the sexual act in the Middle Ages – the matron, the midwife who was required particularly in proceedings for the annulment of marriage. The surgeon Guy of Chauliac describes her as administering aphrodisiacs, making free with her advice and being present 'for a few days' at relations between husband and wife, so as to be in a position to make a report to the doctor who in turn would be able to give his opinion to the judicial authorities. We will be returning to this procedure when we discuss impotence. Furthermore, the matron has more practical knowledge than the doctor, because gynaecological examination tends to take place away from the eyes of men.

There were thus two types of women who had expertise in sexuality: the procuress and the midwife, whom literature often fused into a single figure. Since they had experience, they were necessarily old. Their power enabled them to join the bonds of love, but also to break them. If she helped to give life, the midwife could also belong, like the prostitute, to the world of sterility, distributing contraceptives or abortion-inducing potions. A section of medical discourse – little represented in the works of the most scholarly writers, but in wide circulation nonetheless, especially in the form of the *Secrets of Women* – exploited physiology to rationalize and justify this belief in the wickedness of old women, and thereby echoed the fear that they inspired, and which ran the risk of degenerating into an obsession. In the play by Fernando de Rojas, the heroine was not really a witch; so she did not go to the stake, but was assassinated.

This brief reminder of the sexual content of profane literature puts in the right context the contradictions and dangers which medical discourse had to face. Between religious repression and the most unbridled freedom of life–style, it had a dangerous path to tread. On the one side, profane literature, whether that of Latin comedy or of the fabliaux, reflected a robust health, a debauchery, a licentiousness, that seem to have owed nothing to the refinement of philosophical and scientific thought. These writings often expressed only the elementary satisfaction of a need. One might go on to ask whether simplistic fiction and vulgarity did not mask the functioning of a subtler erotic imagination which took on the appearance of uncouthness and naivety. In this way, remarkably revealing details and additions can be seen slipping through via the tradition of antiquity, and especially of Ovid's tales. Certain of these literary creations merely provided an outlet for the imagination. More skilful were the texts that proposed, in a poetic or philosophical guise, a veritable sexual initiation. There is hardly any doubt but that an art of love, based, as it should be, on secrecy, was in circulation, and that it drew some of its information from medical literature. Andreas Capellanus, one of those 'secretaries' of love, openly took several of his arguments from the 'physicians'. But more than this, his definitions, and the advice he gave, were not only philosophical or Ovidian in inspiration; a great number of them referred to the physiological representation of the mechanisms that govern the sexual act. In this way is raised the question of the responsibility of the doctor, whose voice was heard more and more clearly during the Christian Middle Ages.

THE GROWTH AND DEVELOPMENT OF AN EROTIC SCIENCE

We must return to the medical texts. In any case, we had hardly strayed
from them: literature insistently referred us back to them. That every
discourse on sexuality was first of all a *scientia sexualis* was consistent
with medicine's deep vocation, and our chapter on diseases will show
the necessary and praiseworthy attempts undertaken by medical art to
explain and convince, to cure and even to prevent, as far as its feeble
means permitted. Medicine did not hold entire sway, however; behind
the appearances, we should try to find what it was that enabled medieval
people to maintain a clear separation between their thoughts and actions
– in other words, we must consider what played the part of instigating
or prolonging pleasure, and what contrasts were drawn between pure
'animal' instinct and human behaviour. In plain language, one must
seek, between the lines of medical discourse, the traces of an art of loving
which other kinds of works seem to have been capable of revealing to a
public of initiates at a specific period in the Middle Ages.

The origins of Western sexual knowledge

Our enquiry begins in the eleventh century. As in the other domains of
medical science, it was during this period that the discourse on sexuality
was codified. The following centuries contributed their own additions
and innovations, but these later discoveries were made to fit the basic
pattern laid down by this first rebirth of medical science. The *De Coitu*
ascribed to Constantine the African is of particular importance, because
it was at the origin of the Western tradition;[73] several other works called
De Coitu were more or less revised and enlarged paraphrases of this
model. That is why we will give an outline of it; it will also enable us
to bring together, in a homogeneous textual outline, notions that we
have already set out in dispersed form in the chapter on physiology.

Outline of the treatise of Constantine the African
Introduction   Like all the texts dealing with reproduction the *De Coitu*
includes a preliminary declaration, asserting that the sexual act was
wished by God so that the race should not die out and so that the
pleasure which accompanies it should make it easier to perform.

The operation of the organs   The lines following are devoted to the role
played by the different vital organs and we have discussed the subject

at sufficient length, so there is no need to come back to it. The different cases of impotence and sterility are described.

Definition of the male sperm   Semen is a moist, pure, warm substance; the sperm, according to Galen, is *spiritus* and *humor spumosus*. The humour is foamy because of its movement, as happens in the case of the sea during a tempest. Once it leaves the adequate receptacle, the spirit of the seed evaporates.

The role of the testicles   They receive the semen, transform it and send it to the penis. During this process, the operation of simple qualities modifies the quality and consistency of the semen: essentially, heat makes it subtle, cold makes it thick, dryness makes it moderate and moistness makes it abundant.

Division of hot and cold   Heat is the property of masculinity and cold the prerogative of femininity. The dominant presence of one or other of these qualities in the testicles determines both the appearance of the individual, his or her endowment with hair and the sex of the future embryo, as well as the sexual abilities of the person who will develop out of it. The importance of the testicles in this 'programming' is considerable and the author comes back to the right–left division. The size of the right testicle show an aptitude to beget males, and the opposite is also true. All this is determined at the stage of puberty. One also has to take into account the localization of the sperm in the womb. Thus the combination of an emission coming from a strong right testicle, with a reception in the left part of the womb, produces the effeminate man. The origin of the masculine woman can easily be guessed.

The best conditions for coitus   Galen, referring to Epicurus, declares that no living creature can be in good health if it is in a state of abstinence. The most favourable moment for performing the sexual act is when the body reaches the exact temperament. The sleep that may follow is especially recommended to the woman so that she may more easily retain the semen. Considerations on digestion are very important. The middle of the night is an unfavourable moment, for the food has not undergone a complete coction. In the morning, before any meal, is not the moment to choose, either. The author tries to establish a link between the perfection of the future embryo and the state of digestion.

On the usefulness of coitus for certain people   Reinforced by the authority of Galen and Hippocrates, Constantine asserts that coitus is necessary and beneficial for those in whom the viscous phlegm is abundant, as well as the ardent vapour which, if allowed to infest the temperament of the body, could greatly damage the organism. In this case, coitus is beneficial, for it dissolves the superfluous matter, calms and cools down the body, and acts as a stimulant on it. For temperaments deficient in phlegm or weak in vital force, the sexual act is dangerous. Coitus brings both cold and weakness to man: in those who have a tendency to it, the effect can only be disastrous. The author then moves on to more specific consequences: coitus dissipates the poor state of the body and calms its fury. This last observation is based on the violence of rutting among certain animals. It is beneficial for melancholics and brings the demented back to their senses. It dissipates the lover's desire even if the lover enjoys satisfaction with another woman than the one he desired. This interesting passage will become an integral part of the treatise of Arnald of Villanova on heroic love to which we have already referred.[74]

Indispositions caused by coitus   During coitus, certain men experience contractions and depression. Others are afflicted by tremblings, by an unpleasant odour, or by a swelling of the stomach. Yet others are affected by piercing noises, or suffer headaches after coitus. As a preliminary to any explanation, Constantine recalls that sperm is constituted by the very essence of the healthy parts of the body and that what is emitted is not only humour, but also the vital spirit transmitted with the semen by way of the arteries. In these circumstances the weakness which follows the sexual act should be no cause for surprise. Coitus may be fatal and Galen declared that animals that indulge in intense sexual activity die more rapidly than the others. It can equally be observed that eunuchs live longer. Taking his information from different sources, Constantine gives an explanation for these different maladies. Contractions and melancholy are explained by a pernicious change in the humours, caused by the heat that attends the moment of coitus; indeed, the body stiffens when it contains harmful humours and if it has suddenly been heated. Depression occurs at the same time. The trembling is explained by the noxious effect of superfluous matter. The unpleasant odour comes about from the putrid mixture which is dissolved at the moment of coitus by the heat produced during the act. The swelling of the stomach happens when the warm quality is weak, and this is caused by the melancholy humour. This last inconvenience is experienced by those who always have a desire for coitus, because of the great quantity of 'flatulence' set in motion during the act; and because the flatulence is contained in the organism, noises and rumblings in the belly are produced; for certain men, a high-pitched noise is heard coming from outside. In other individuals the headache is caused by the rising of the vapour formed from the moistness of the blood as well as the superfluous matter in the body; the cephalalgia also comes from a bad mixture of humours.

Before continuing with our analysis of the treatise, we should recognize that Constantine is led by a desire for method and that, although they are hardly satisfying to us, the explanations do their utmost to leave nothing unclarified. Doubtless, his work includes a number of correct observations. The analogy of coitus with an intense muscular effort explains part of the phenomena noted; the setting into motion of the mass of the intestines, post-coital sweating, the stimulation of secretions and disturbances caused by an increase in arterial tension and by the acceleration of the cardiac rhythm.

Aphrodisiacs and anaphrodisiacs   The last and longest part of the treatise is devoted to aphrodisiacs and anaphrodisiacs. The pharmacopoeia that the author bases his discussion on is arranged in the following way. First dealt with are the products that give rise to semen and those that destroy it, those that stimulate desire and those that prevent it; the treatise then discusses types of foods which, on the contrary, dry up and diminish the quantity of semen, if individuals abstain from products of the opposite kind.

The ingredients that contain nourishment and that cause flatulence reproduce the very essence of semen, since it is established that its nature is to be both humour and breath. A third faculty can be attributed to foodstuffs – that of being able to attract by virtue of their heat. These three faculties are all found together in the chickpea; it is very nourishing, causes flatulence and is hot and moist in quality. By itself, it is capable of being turned into semen. There follows a long section on the products capable of engendering semen and extracting it from the organism. From among them, we will cite the satyrion (also called 'man orchis' or 'fox's testicle') which inflames desire. Dioscorides adds that its root, when held in the hand, stimulates the amorous appetite.

On the other hand, anaphrodisiacs are foodstuffs that do not engender semen, but dry it and dissolve flatulence – in other words, all the products that are hot and dry. Likewise, cold foodstuffs are the enemies of semen, for they thicken it and extinguish desire. An additional remark points out that cold and dry products restrain the sexual appetite. The end of the text is devoted to new aphrodisiacs, among which one can find three ointments applied to the genital organs or onto the soles of the feet.

At the end of this rapid review of the *De Coitu* of Constantine, several remarks are called for. With the possible exception of indications relating to the moment of coitus, the treatise hardly provides us with any knowledge that is not contained in another part of the compiler's work.[75]

In reality, it does not manage to express anything other than the way sexuality became a province of medicine. The conception of the human body as being governed by the laws of physics and mechanics leaves little place for discussions of anything not pathological. The sexual act is linked only to physiological equilibrium. This whole treatise is an encouragement to the taking of medicine, if one excepts the relatively unimportant practice of applying ointments. It is, for the time being, the only gap through which there slips the shadow of an erotic art, in the context of a work which as a whole does not manage to speak even once about the female partner.

To counterbalance this, we will give a brief analysis of another treatise called *De Coitu* which, attributed to Maimonides,[76] was circulating from the end of the twelfth century onwards in Western Jewish communities and was translated into Latin by John of Capua a hundred years later. The *De Coitu* of Maimonides conveys the same scientific information as that we have already had occasion to meet with. Nonetheless, it presents a major difference: it establishes a link between the mind and the mechanism of the body. When coitus is desired, man experiences a perfect erection, whereas in the opposite case the male member is weak and dry. Thus monks who wish to maintain their chastity more easily must remember that they owe their lack of guilty thoughts to their lack of erotic activity. If it is performed without any great desire, the sexual act is frustrating, and carried out with a person who is not desired, it is not recommended – such partners include, for instance, a virgin, an old woman, a girl who has not reached maturity, a partner who has had no sexual activity for several years, or one who is menstruating or ill. The advice on diet and pharmacopoeia is established in accordance with Galenic principles and tells us nothing new. The same importance is granted to attacks of flatulence. A preoccupation which has to do with the art of love – a preoccupation which has left hardly any trace in Western texts of the period – occupies a major place in the treatise. We are referring to ointments, and especially to a product with which one smears the glans penis three hours before coitus and which enables an erection to be maintained after a first ejaculation. One can thus satisfy a partner who takes a long time to be stimulated, or even several women. Finally, we will mention a practice which consists in oiling the region of the coccyx and keeping it warm, especially the anal sphincter and the upper part of the thighs, so as to make it easier for the warmth of the blood to reach the genital organs. We are already acquainted with the importance of these different factors. These few remarks allow us to grasp the difference there is between the two treatises under consideration. In the second, psychology plays its part and the sexual act is no longer a discharge of humours. Furthermore, sexuality is not considered merely from the standpoint of pathology: a technique whereby the male may control his desire finds a place in it.

Before exploring in greater detail the Arab erotic treatises, we must refer to a true 'feminist' text, even if it seems that it was written by male authors.[77] The set of the three treatises attributed to the prestigious lady of Salerno, Trotula, includes an interesting prologue. It cannot be denied that the medicine of the age ascribed to women certain characteristics and a certain physiological make-up that created a formidable set of constraints. The author's desire to come to the help of the victims of this condition lacks neither originality nor even grandeur. Let the reader judge: 'Since, then, women are by nature weaker than men, it follows that diseases are more frequent among them. And when diseases occur in their private parts, because of their sense of shame and the fragility of their condition, they dare not themselves reveal the sufferings of their illness to the doctor.' It would indeed be difficult to imagine a situation in which the barrier between the two sexes, and between women and any help that science might bring, was more insurmountable then here.

As we have already pointed out, Trotula pays particular attention to the purgation that happens in women by means of the menses. The diseases and more especially the infections that follow childbirth find a place in this treatise. Trotula mentions two accidents which may occur after coitus: a swelling of the vulva, as well as a slipping down of the womb caused by the excessive size of the virile organ. One also finds the recipe for a remedy capable of giving back the appearance of virginity, but the reason is honourable, since the dilation of the vagina could prevent conception. Trotula then registers the physiological disturbances that can occur in widows or in people who have taken a vow of chastity. The delicacy with which specifically gynaecological problems are treated, as well as the understanding shown towards the female condition, are such exceptional attitudes in the Middle Ages that we cannot avoid mentioning them.

A treatise on the care to be taken of the body and the face takes its cue from medical preoccupations. In its diversity, it has nothing to envy our modern advice on beauty care. In it, one finds products to make the face white or more highly coloured, and ways of smoothing out wrinkles, skin spots and facial veins. Here is something for sunburn: 'Let us note a special ointment, effective against sunburn and against any sort of chapped skin, especially that caused by the wind, and against pimples on the face caused by the air: this ointment, which is used by the ladies of Salerno, also works for marks and scratches.' We learn that another ointment also enables one to get rid of the pustules of the plague. Superfluous facial hair, scabies, various types of dermatosis, and many other afflictions which disfigure a woman's beauty find a remedy. There are recipes for whitening black teeth, keeping the gums healthy and making the breath smell sweet, especially using this procedure:

I for my part have seen a Saracen woman with the help of this medicine rid a great number of women of bad breath; she took several laurel leaves, a little musk, and ordered the women to hold it under their tongues so as to render the heaviness of their breath unnoticeable. I also recommend this: the woman should hold this remedy under her tongue day and night, especially when she has to carry out the sexual act with someone.

The end of the treatise is devoted to hairdressing, and in it can be found ways of dyeing hair, of making it grow very long and turning it black, following a 'Saracen procedure'. At this point we will bring to a close our discussion of this art of decoration on which Arab influence can be sensed. Doubtless, beauty care was influenced by the Muslim world and by the refinements of the harem. This knowledge constituted a prelude to the act of love and must be linked to the knowledge of erotic relations between man and woman. The influence of Ovid and the oriental texts could have led to a positive evaluation, a poeticization of the art of embellishing women. Nothing of the sort happened. For preachers, hair was evil and attracted the demons prowling around human beings. Rouge, and any care a woman might take over her appearance in fact constituted a formidable attempt at dissimulation.[78] Thanks to these artifices, men were incapable, even when armed with the weapon of physiognomy, of discovering a woman's real temperament: they thus exposed themselves to a number of mistaken judgements and disappointments. Latin poetry echoed this mystification, at the same time as it demonstrated the way precepts transmitted by the *De Ornatu Mulierum* were put into practice:

> Altera jejunat mense, minuitque cruorem,
>   Ut prorsus quare palleat ipsa facit.
> Nam quae non pallet sibi rustica quaeque videtur;
>   Hic decet, hic color est verus amantis, ait.
> Haec quoque diversis sua sordibus inficit ora;
>   Sed quare melior quaeritur arte color?[79]

East and West: a different language

So as to draw up a few points of comparison, it is necessary to investigate the Arab erotic texts; on the basis of identical information, the East in fact took to a logical conclusion an art of love which found a home in the West only by accident. Based originally on the work of Rufus of Ephesus, but also influenced by Indian sources, the Arab treatises devoted to sexual hygiene were part of a long tradition: between the ninth and thirteenth centuries, one may suppose the existence of some hundred works of the same type.[80] These texts met with particular favour from princes, for whom they were most often written. From the same point of view one may cite, for medieval France, the *Dialogue de Placides et Timéo* which is an example of a form of teaching relating to the secrets of women and designed for the initiation of a prince, with the sole difference that it is mostly limited to gynaecology.[81] In the Arab world, there were few doctors or scholars who did not indulge in the genre of the treatise on sexual hygiene; one of the characteristics of these writings is that they mix in with the didactic part stories or anecdotes which themselves aim at stimulating erotic thoughts.

Next to the *De Coitu* of Constantine the African, one must set the analogous treatises of Qusṭâ ibn Lûqâ (*c.* 820–912), who likewise was a translator of numerous medical texts, in this case from Greek into Arabic.[82] Both writers were clearly inspired by Galen, Qusṭâ ibn Lûqâ recommends that the sexual act should be performed in accordance with the age of the person concerned, the seasons of the year, and the time of day relative to digestion. We will restrict ourselves to picking out the ideas that we think should figure in a history of sexuality. For the author, a desire for coitus and an erection that are stimulated by the imagination or by a movement of the soul are not beneficial: after the act, furthermore, the man is tired, worried and depressed. On the other hand, coitus when it is demanded by the organism makes him lively and cheerful. As for the vertical position, it is one of the least suitable in which to perform the act, since it is tiring and harms the organs. This observation also applies to pederasty.

In the man, a close connection is established between the reproductive organs and the chest, so that sexual excess leads to blood being expectorated, and a whole series of observations come to back up this assertion. At puberty, changes occur in the chest and, after castration, servants have a higher voice and bigger breasts. The link between the head, the breasts and the genitals is well established in women. Caresses and bites during love-play, in human beings as well as in animals, demonstrate the relationship that exists between the head and the genitals. Sexual relations are highly recommended, since the sperm retained in the body goes rotten and from it there comes a deadly emanation. Qusṭâ ibn Lûqâ goes on to refer to the cataleptic state of women in thrall to an epileptic fit, and concludes his treatise with a discussion of priapism.

One of the most revealing texts is the *Book of Conversation with Friends on the Intimate Relations Between Lovers in the Domain of the Science of Sexuality*, written by a Jew converted to Islam, as-Samau'al ibn Yahyâ (died 1180).[83] As its title leads us to imagine in advance, it is also one of the most precise as regards the application of scientific and medical thought to sexuality. Right from the start, it tackles the subject of homosexuality and heterosexuality:

Know that many eminent men of our time indulge in the frivolous practice of having relations with boys. Several of them have been led to do so by their doctors, who have persuaded them that sexual union with women leads more quickly to old age and to the weakness of age, and causes podagra and haemorrhoids, whereas relations with boys are less harmful.

The rest of the treatise is devoted to a long and highly technical comparison between the anal sphincter and the muscles of the uterus. For the performance of coitus with a woman, the use of an astringent has its inconveniences, for its action is incomplete. The astringent acts on the cervix of the uterus, but not on the section circumscribed by the labia. In the case of homosexuality the man is in fuller control of his anal sphincter than the woman of her vagina.

The second chapter is devoted to lesbians, who in many cases indulge in lesbian activities only occasionally, for their sexual behaviour is explained by the difficulty they have in reaching orgasm with their male partner. Others adopt this behaviour because of the smallness of their cervix, which leads to difficulties in intercourse; others too because they reach a climax too quickly, whereas their partner is too slow; and yet others because of illnesses or wounds. These latter, the author adds, are suitable only for an impotent man, for during intercourse only their clitoris plays any part. One can thus see that he clearly dissociates clitoral pleasure from vaginal orgasm.

We will excerpt from this treatise the extraordinary portrait of very modern women:

There is a certain category of women who surpass others in intelligence and subtility. There is a great deal of the masculine in their nature, to such an extent that in their movements, and in the tone of their voice, they bear a certain resemblance to men. They also like being the active partners. A women like this is capable of vanquishing the man who lets her. When her desire is aroused, she does not shrink from seduction. When she has no desire, then she is not ready for sexual intercourse. This places her in a delicate situation with regard to the desires of men and leads her to Sapphism. One has to look for the majority of those who possess these qualities amongst the elegant women, those capable of writing and reciting – amongst the cultivated women.

After this astonishing portrait, drawn in the thirteenth century, the treatise concludes with the reasons that lead people to live together. The metaphors show that in a certain system of thought the erotic function can be reconciled with the spiritual impulse:

The third reason in favour of the existence of coitus is that it has the advantage of unbinding the shackles constituted by the body, and that it allows us to pass across the walls of one's prison and to set the stones of the prison into motion, so as to make it possible for the soul to take its flight from this prison.

The Prophet clearly indicated, moreover, that women are the best way the initiated man can occupy his time in this world. The philosophy of Islam, which constituted a particularly favourable factor in medical, technical and psychological research into the performance of the sexual act, can easily be seen at work here. Even if the context may still be foreign to us, the fact remains that thinking in this field is on a par with the best ideas of modern psychology. From another part of the treatise, more especially devoted to pharmacopoeia, we glean the assertion that a stay in places like Mecca, Medina or Baghdad allows virile potency to be reinforced.

At-Tîfâshî (1184–1253), in his book *On What is Useful to Women and Men in the Practice of Sexual Intercourse, on What is Favourable, and What is Harmful*, weaves together precepts, stories and pharmacopoeia.[84] From his teaching, we will mention the way he establishes a classification of women determined by their sexual behaviour: in the East as in the West, physiognomy has an influence on all psychological analysis. One comes across some rather blunt formulations: what has to be sought in a woman is tightness, heat and dryness. In the absence of these three qualities, self-satisfaction is preferable and leads to a more pleasant ejaculation. One can also read a remark which Western doctors accepted only much later and more hesitantly:

If you want to arouse the girl, play with her breasts and you will witness a marvel, for her seminal flux is just below the collar bones, which are connected to her breasts, as the testicles are to the penis. . . . The woman is drawn to sexual commerce, when one plays with her breasts, and, especially when she is older, do not deprive her of this pleasure.

In scientific tests it is repeated time and time again that there is a link between menstrual blood and the womb and that menstrual blood is transformed into milk to feed the child with, but this is the first time we have encountered an explanation like this one for the sensitivity of the breasts. The text of at-Tîfâshî establishes a physiological link between the seminal flux and the arousal of the breasts, but he does not specify the details of the mechanism. The tone of the recommendation is what is striking – it does not appear at all to be a commonplace.

Ibn Falîta, a writer of the first half of the fourteenth century, adds some new details.[85] His work contains numerous erotic stories which all have a didactic value. The man has to be careful that he does not put his partner off with his body odour, his breath or his sweat. How ridiculous are those men who dye their hair in order to appear younger, once the moment of truth arrives! Ibn Falîta denounces male behaviour which deprives the woman of her pleasure, such as '*ejaculatio praecox*' or premature withdrawal of the male member. Tugging on the hair at the moment of orgasm appears as a way of intensifying pleasure. In regard to this, the author undertakes a minutely detailed study of the positions which allow this to be carried out. This assertion proves that the different positions were a suitable subject for those writers who devoted themselves to eroticism. An anecdote recounted by this same Ibn Falîta seems to confirm this. A man reads in the Koran the story of Joseph and his host's wife. After Joseph's refusal to give in to the advances of the temptress, the reader closes the book and says to himself: 'My God, if at least you had led me to her, I would have shown her ways of performing coitus that she does not know and which she has never even heard of.' The story is an amusing one, and the way the character in it thinks presupposes the existence of an erotic art based on various different positions. It is probable that this art was able, by more or less secret paths, to infiltrate the Christian world: certain cultural areas were propitious to influences between East and West. The Catalan manuscript from the fourteenth century that we analyse at the end of this chapter is proof positive of this possibility: the art of loving was this time clearly transcribed, without any recourse to rhetorical camouflage, and it was placed firmly within a medical context.

The Arabic treatises that we have just briefly analysed have the peculiar feature of being written more often than not by doctors, but for non-doctors to read; they tend to spread information – one might say in certain cases a kind of initiation – by placing their science in the service of an art of living. The Western genre which might at first sight seem to be close to them is that of the countless *Secrets*, the aim of which seems similar; an author, who as often as not says he is a cleric, claims to be revealing to a lay person (or to another cleric who will pass on his knowledge) the way one should behave with women. The ground covered is neither that of theology, nor that of ethics, and the arguments are drawn entirely from scientific thought, a thought based on the same sources as Arabic medicine. As one of the descendants of this literature, the *Dialogue de Placides et Timéo*, bears witness, the class of intellectuals had to appropriate knowledge on sexuality so as to insert it into a discourse of a political nature.

At the origin of this kind of text is an Arabic version, the *Sirr-al-'asrâr* known in Latin under the title of *Secretum Secretorum*.[86] This work immediately met with a great success in Europe: it was translated from Arabic into Persian, into Castilian, into Hebrew twice and into Latin. From Latin there were several translations into Italian, French and English, while a Hebrew text gave rise to a version in the Russian of Smolensk. The *Secret of Secrets* is linked to the epistolary genre thanks to the insertion of a so-called 'Letter of Aristotle to Alexander'. This form makes the work capable of bringing together all the scientific advice that the expert can give to the prince; and so, quite naturally, it includes a health regimen, as well as a treatise on physiognomy. Certain Arabic versions did not forget to give room to the 'precautions that a king must take so as not to fall victim to the women of the harem – or to his doctors'. In fact, what we have here is another treatise of scientific popularization, in a form which, whatever the field envisaged may be, allow the person in possession of knowledge – or pseudo-knowledge – to be placed in a situation of domination with respect to the person in possession of political power. This literary mechanism, which is that of a narrative fiction, in other words a dialogue between a venerable philosopher and his disciple, functions in reality as a message from author to author, with the added effectiveness created by didactic scruple and the tone of confidentiality. The knowledge transmitted is often of little value, even by the standards of medieval thought, but the fields of enquiry are well chosen. One is physiognomy, which enables the temperament of each individual to be known, in other words the thoughts and ulterior motives of one's closest companions to be guessed and predicted. The other is the field of medicine. Both are based on the perspicacity that the intellectual ascribes to himself. Geomancy and astrology are two other components of this faculty. Within the field of medicine, sexuality and the knowledge of women are guaranteed to arouse interest. That is why the tradition of the *Secretum Secretorum* involved handing on information in a highly secretive way, as if it were esoteric knowledge. As an example of the kind of tone in which such works were written, here is a passage from a later avatar of the original model, in which the master is addressing his royal disciple: the woman appropriates the man's heat during coitus, and the more she feels this heat, the more she desires it. At this moment the master breaks off from his discourse to declare;

In this field, when you have been more obedient, when I have received from you more marks of affection, then I will tell you the deep secrets that must not be revealed to anyone, except to one's dearest friend: it is the very essence of the secrets of nature. And philosophers say that you must only write in a small, fine, slender characters that are hard to decipher, and on poor quality parchment that is difficult to read and does not last long, in sentences whose meaning is hidden, because a knowledge freely transmitted and made public is of little value, whereas, if it is discovered with difficulty, it is rich in meaning, and precious.[87]

The writer here creates the extraordinary fiction that he hopes the material support of his creation will be destroyed (in fact his work is preserved in two splendid manuscripts). One could go on at great length about this emotional blackmail, this relationship between friends, presupposed by pedagogical practice. It would be wrong, in our opinion, to see in this strategy nothing but an advertising ploy.

Another tradition of *Secrets* developed and spread: that of the *Secrets of Women*, which was to have a great influence, not only on that public that was capable of reading, but also on the thinking of ordinary people. In fact, these texts are never erotic treatises. They always follow the same pattern and there is always a certain deceit about the character of the merchandise being purveyed. The work contains, apart from the everlasting information about the nature of sperm and the menses, a treatise on embryology which follows the development of the foetus month by month. The disciple who receives such an education will hardly draw any profit from it in his relations with women and he will perhaps have the painful feeling that the formation of the embryo is not entirely determined by the mixing of the two seeds. The *Secrets of Women* transmit an Aristotelian mode of thought that is so bluntly expressed that it becomes caricatural. Menses are a purgation; their function is to feed the embryo and, without any physiological clarification being given, we are told that they have a part to play during coitus, in the following way:

When a woman has intercourse with a man, because of the intensity of her pleasure (since the man's erect penis rubs against her aroused nerves and veins and sets them in motion), then the vulva swells and emits menstrual blood and this is the sexual act considered as natural, because it is the natural way of performing coitus.[88]

This is the kind of sexual education given and it is difficult to overlook the forcefulness of the dogmatic assertion contained in the term 'natural'. Female pleasure is totally subordinated to the phallus by means of an infallible physiological mechanism. The last part of the sentence seems to exclude any attempt at thinking about different positions or any other components of female pleasure. This text used to be ascribed to Albert the Great: this is no longer held to be so.[89] Indeed, such an ascription was a slander on the scientific mind of the man who, in the thirteenth century, showed the greatest indulgence for, and the greatest understanding of, women.

In the later versions of the text, the introduction can be extreme in its brutality. Let the reader judge:

That friend to whom he [Albert the Great] writes was a Priest, who begged him to write the *Book of the Secrets of Women*, for the reason that women are venomous during the time of their flowers and so very dangerous that they poison beasts with their glance and little children in their cots, sully and stain mirrors, and on some occasions those men who lie with them in carnal intercourse are made leprous.[90]

All these assertions do indeed appear either in the *De Anima*, or in Albert's paraphrase of the *De Animalibus*. The underlying idea is the one we have mentioned in studying the phenomenon of menstruation: the woman secretes a poison against which she herself is immunized. This idea played only a contingent part in Albert's work. On the other hand, those who appeal to his authority have picked out very carefully the examples and assertions that are the least favourable to women. It is these texts that left a deep mark on popular sensibility for the three centuries that followed. In the second half of the thirteenth century, clerics saw their political and economic functions multiplying as a result of their mastery of writing. These intellectuals – whether marginal or not, either confined to celibacy or else indulging in more or less abnormal sexual behaviour – did not understand the problems of the couple. They were more at their ease when it came to a discourse that attempted to play a preventive role in warding off the threats of existence. In order to reassure, one had first to scare. Epidemics often led to scapegoats being sought out, but this was a phenomenon of such importance that it could alert the collective sensitivities of a whole people. A health regimen was better adapted to a discourse from individual to individual, which explains its extraordinary success. As far as relations with women – and the sexual act – are concerned, this is the main subject, the terrain on which in every time and in every place the ancestral fears of the male have focused. The exploitation of the theme that originated in the thirteenth century was to be so successful that it gave birth to what might be called the great fear of women, intimately linked to all those sadistic elements that were to be unleashed at the time of the rise and subsequent repression of witchcraft. Jean Delumeau is right to say that women were seen as the instigators of calamity.[91]

The Secrets of Women genre thus does not fit in with the topic of our research. The reason we have chosen to present it after the Arab treatises on sexual hygiene is that, of these two genres, the one constitutes an opening onto the psychological dimension of love, whereas the other contributes to maintaining the difference between the two sexes; and yet their titles could have allowed one to presage identical functions.

Eroticism regained

Although the preservation of some Latin translation or other does not enable us to presuppose a direct acquaintance with the Arab treatises we have just discussed, most of the medical sources that the latter were based on were known in the West. First and foremost there was the *Canon* of Avicenna which was in the widest circulation in university and medical circles from the earliest years of the thirteenth century onwards. In it, the usefulness of coitus for keeping oneself physically and mentally healthy was reasserted; in the same period, the fear of death that a woman might feel at the idea of another pregnancy was included among the arguments put forward in favour of contraception. But above all, through Avicenna's influence, the consensus of doctors regarding the existence of female sperm was maintained; even in authors who granted it only a restricted role, the link between fecundity and a female emission was preserved. This enabled medical preoccupations to be displaced and created the possibility of making room for the beginnings of an erotic art: the quest for pleasure as such was to be one of the components of every discourse on the conditions for performing the sexual act, since in the background loomed the necessity of encouraging conception. Sterility and frigidity became synonymous. All of these factors led to the notion of sin being forgotten when this was necessary, and to the possibility of a freedom in thought and behaviour which might have struck us as irreversible, were it not for the opposing current transmitted by the Aristotelian vision of woman and the failure of the centuries that followed the Middle Ages to follow up this vision of freedom.

The necessity for the man and the woman to reach a simultaneous emission, and so to share their pleasure, gave doctors an 'excuse' for tackling certain problems:

There is no shame in the doctor speaking about the increase in the size of the penis, or the narrowing of the receptive organs, or of the woman's pleasure, since these are factors that play a part in reproduction. For the small size of the penis is often a hindrance to climax and emission in the woman. Now, when the woman does not emit any sperm, conception cannot take place.[92]

Avicenna thus granted doctors freedom of expression with, in addition, a justification that could not be neglected in a society that made sexual intercourse serve the sole end of reproduction. Medieval writers backed Avicenna's assertion with an aphorism from another author writing in Arabic: 'You must not be ashamed to ask the patient about everything.'[93] Doctors made use of this freedom of expression first of all in what seemed to be most within their field of competence, that is, in everything that had to do with the structure of the genitals and any incompatibilities that might arise between them – and so one comes across long lists of ointments, pessaries or fumigations designed to make the respective organs bigger or smaller, wider or narrower or stronger and so on. The authors often echoed currently observed contemporary practices but they sometimes made them serve a purpose for which they were doubtless not intended. Here, for example, are the words of a fifteenth-century commentator on Avicenna:

I have heard that the opulent women in the Kingdom of Navarre had recourse to vaginal fumigations when they were about to unite with their husbands. The reason perhaps is that this makes conception easier by bringing the womb down somewhat and enabling the male sperm to be completely taken in, but these women should afterwards breathe in spices through the nose so that once the sperm has been taken in, the womb may rise again.[94]

Doctors thus accepted without hesitation all of Avicenna's prescriptions meant to make sexual intercourse easier or more pleasant; they even added a certain amount of advice drawn from other sources or from their own experience, but they did not always manage to do so without seeing them as a means of obtaining greater fecundity. With this type of prescription, we are still dealing with a sexuality in thrall to a purely medical approach. Another short paragraph from the *Canon* of Avicenna risked creating a greater difficulty, since its advice lay completely outside the framework of pharmacopoeia. Indeed, in the chapter devoted to curing sterility, the foreplay that should precede and accompany the sexual act is discussed:

Men should take their time over playing with women who do not have a poor complexion. They should caress their breasts and pubis, and enfold their partners in their arms without really performing the act. And when their desire is fully roused, they should unite with the woman, rubbing the area between the anus and the vulva. For this is the seat of pleasure. They should watch out for the moment when the woman clings more tightly, when her eyes start to go red, her breathing becomes more rapid and she starts to stammer.[95]

We still have to ask ourselves what the future of such advice within Western medieval science was to be. If one refers to the first commentator who studied, at the beginning of the fourteenth century, the third book of the *Canon*, one has to conclude that this paragraph was in circulation only as a dead letter. Indeed, Gentile da Foligno abstained from commenting on the description of foreplay and was interested merely in defining the meaning of 'a complexion that is not poor', an expression that he glossed as 'which is not too strong in heat'.[96] In the fifteenth century, Jacques Despars contented himself with paraphrasing the passage in question as a whole and with pointing out that the end of the description corresponds to the moment when the woman is on the point of emitting her seed; the purpose of the performance was thus safeguarded.[97] If one refers, however, to medical works which are not direct commentaries on the *Canon*, but bear the mark of its influence, one has to note that the theme was taken up and considerably expanded. In the *Lily of Medicine*, begun in 1303, Bernard of Gordon repeats almost verbatim Avicenna's advice *ad excitandum feminam*.[98] Several years later, John of Gaddesden in the *Rosa Anglica* waxed more eloquent. His work is presented as a medical summa. A great deal of space is taken up by the remedies deemed to palliate different anatomical incompatibilities, but also by contraceptive procedures. John of Gaddesden takes up Avicenna's description of foreplay, adding several details which show that he is taking into consideration the woman's erogenous zones, although no reference is made to the specific role played by the clitoris.[99] What is more surprising is that he envisages the woman's taking the initiative if the first attempt has met with failure, that is, if the woman has not experienced any pleasure and her partner is showing signs of difficulty in managing a second attempt. So, he recommends that the man lie on his back (a position which makes erection easier) and that the woman performs a series of movements designed to arouse male desire.[100] We do not know in what position the act is to be performed. This text contains one of the rare examples to show the woman as capable of taking the initiative, something that presupposes her knowing the right technique. The art of the kiss is also discussed in detail, following a tradition we have already encountered in Trotula. Apart from presupposing the learning of an erotic technique, John of Gaddesden pays particular attention to mental processes. From the lengthy discussion devoted to treating male 'sterility', we will quote the following:

The fourth condition necessary is that the troubles of the soul be removed, the way of begetting children taught and everything that works against it should be shunned. Men should amuse themselves, they should listen to bawdy songs, they should about the sexual act and watch other men or animals performing it. They should fly from sadness by all means possible, by looking at beautiful young girls or engaging in conversation with them. They should sit next to the fire to warm their bellies. They must completely avoid anger and sorrow.[101]

The eroticization of medical discourse that we find in John of Gaddesden is no accident; for the simple reason that texts drew mutually on one another, it actually increased. In the fifteenth century, Michael Savonarola introduced the chapter on ways of making conception easier with these words: 'I will not hesitate to describe what is useful for procreation, even if it does not seem decent.'[102] The chapter opens with considerations on the desirable frequency of coitus: the frequency cannot be dealt wth in generalizations, but has to be adapted to the age, complexion, regimen and habits of the individual. Michael Savonarola relies on the 'doctor's prudence' for 'what he cannot set down in writing.' The practitioner's role is not limited to composing treatises and we can guess at the importance of that which could not be entrusted to writing, but only said in the secrecy of the medical consultation. At about the same period, Jacques Despars pointed out that a doctor could acquire glory and fortune by dispensing this type of advice to his illustrious patients.[103] This is an element that must not be neglected in a history of attitudes and ideas.

Michael Savonarola takes up Avicenna's description of foreplay, adding as he does so a few details, for example about 'the tips of the breasts'.[104] He pays special attention to the way the act proceeds and counsels that all haste should be avoided, but pleasure prolonged. In addition, we again come across the importance granted to the distractions of the spirit and the stimulation of the imagination:

Men should listen to stories, songs and other similar things which lead to love, they should look at very beautiful ladies, and imagine the sexual act at great length until they feel fortified. Then, after many other things about which I will say nothing because they are known by everyone, they should move on to the act with a good-tempered, well-educated and pleasant woman, who is neither in the middle of her period nor pregnant. Coitus with a virgin is very tiring.[105]

In this advice meant to cure sterility, what has become of conjugal relations? By the end of the Middle Ages, medicine had placed itself staunchly on the side of the individual. Despite the many things that were not mentioned, Avicenna's lesson had been noted and Arab erotic lore was apparently becoming a component of the didactic message of certain doctors. Ovid was equally omnipresent. The question of female sperm provided an opportunity of affirming the benefits of shared pleasure, in a context that went beyond that of literature meant for doctors alone. Despite the theoretical hesitations that we have pointed out, a text destined for a much wider audience like the *Conciliator* of Pietro d'Abano emphasised the importance of female pleasure; he was even one of the few authors to refer to clitoral sensitivity, though without recognizing with any great degree of precision the organ responsible for it.[106] Concurrently, the favour that the theme of heroic love met with supplied doctors with the theoretical data necessary to make the imagination an integral part of their representation of sexuality. So if one or other writer did not include any information on erotic practices, this was the result of a personal choice rather than of any lack of information.

One of the points on which doctors remained more intransigent concerned the art of sexual positions. Avicenna, moreover, placed 'bad positions' among the dangers of coitus.[107] Most of the time, authors made no allusion to the problem, or, if they did mention it, it was in order to declare that they were not going to say anything about it. Thus Bernard de Gordon: 'Avicenna records many illicit and indecent ways of performing coitus; he tells us which are to be avoided, but since [in the *Canon*] one comes across several shameful remarks, I will leave this subject to one side.'[108] John of Gaddesden remained elliptical on this point, as we have emphasized. Michael Savonarola stuck to those arguments that were least favourable to any variation in position: 'The man should stay above the woman, and not the other way round, so as to prevent the female sperm falling into the penis, and so that he will not become tired; the standing position is very exhausting, which is why it is not allowed.'[109] This disapproval was not motivated by any obstacle such a position might place in the way of conception, but by health concerns and also by the ever-underlying fear that the man may be contaminated by contact with the female seed. Even if religious or moral reasons were not invoked, the condemnation was still implacable; coitus must be performed *in debitis vasis* and *cum suis instrumentis*.

An amusing story, taken from imaginative literature, allows us to surmise that, to the minds of ordinary people, the inverting of above and below created another possible danger. Roberto Zapperi quotes a German verse-story from the fourteenth century. In 'The Monk's Punishment', we witness the sexual initiation of a young cleric by an experienced woman; the man's total ignorance leads to the woman's taking the initiative and exchanging positions. At dawn, the worried monk questions his servant:

'I have often heard that when a man and a woman have been together, children are born. But tell me, by your faith, which of the two bears the child?' – 'I will tell you everything,' replied the servant. 'It's the one underneath.' – 'Woe is me,' thought the monk, who was starting to realise the extent of his misfortune. 'Alas,' he said to himself, 'whatever can I do? What a disaster! *I* was the one underneath. I'm going to have a baby! My honour is lost. And if the abbot notices, how will I live? For the monks will run me out. Death is better than their contempt.'[110]

Clearly, doctors did not adduce this risk when they recommended that the woman stay underneath. But it is easy to imagine that, in a poorly informed milieu, the spread of the idea of the 'female' sperm may have given credit to this link between an exchange of positions and a really monstrous and unnatural phenomenon. Behind medical warnings perceived in only fragmentary fashion, certain people may have espied a risk closely linked to the fantasy of the pregnant man.

On the question of positional variations, doctors in the fourteenth and fifteenth centuries demonstrated a greater intolerance than did Albert the Great, when he accepted the possibility of divergences from the so-called natural method, should conception be prevented *per accidens*, for example in the case of obesity. The theologian established several degrees to the fault committed: 'The slightest deviation is the lateral position, then comes the sitting position, then standing, and finally the greatest is *retrorsum*, like mares. That is why certain people have said of this latter position that it constitutes a mortal sin, but that is not my opinion, as I will make clear elsewhere.'[111]

This indulgence has to be contrasted with the silence of later doctors, who also adopted a cautious attitude with regard to *coitus interruptus*: whereas Avicenna pointed out the danger of 'testicular narrowing', fourteenth- and fifteenth-century writers ascribed to it, more seriously, sacrilegious or heretical context. Without the treatise of Andreas Capellanus doubtless had something to do with this fact. This clearly shows the limits that medical discourse imposed on itself: while it was prepared to find room for, and justify, certain erotic techniques (even outside the framework of marriage), in the name of a necessary preservation of physical and mental health, it could not stray from the 'path of nature'. The techniques which included an art of positions or a way of controlling desire were more often than not tainted by a subversive, sacrilegious or heretical context. Without the treatise of Andreas Capellanus or of others like him, medieval medicine would have been able to speak another language – Arab science gave it the means to do so.

This other language can be found in an astonishing treatise discovered by Guy Beaujouan some years ago. Written in the Catalan language of the fourteenth century, it bears an evocative title: *Speculum al foderi* (literally, *A Mirror for Fuckers*).[113] It met with a strange fate. After the treatise had fallen into oblivion for long centuries, Miquel y Planas prepared an edition of it, but his work did not get any further than the printing proofs.[114] What we have here, however, does not seem to be a dead text: some sheets contained in a manuscript other than the one that preserves it in its entirety bear witness to its circulation.[115]

Right from the start, the author places himself under Arab patronage, opening his treatise with the words 'Albafumet says that . . .'. Today, nothing authorizes us to consider it as being really a translation rather

than an adaptation or an original work. There is no doubt that it bears
the trace of Eastern influences – all the more so since Catalonia was a
privileged place for exchanges between different cultures. Although it
has numerous points in common with the genre of Western treatises *de
coitu*, its contents make it closer to the Arabic treatises on sexual health
care that we have mentioned. Like them, it does not address itself merely
to doctors, but aims at reaching a wider public,[116] perhaps even one
including women; as well as giving advice meant for both sexes, the text
is preceded or followed in the manuscript by a Catalan adaptation of
Trotula and a treatise on beauty care.

In the way it begins, the *Speculum* differs hardly at all from the *de coitu*
treatises in Latin. Like them, it adopts a strictly medical point of view.
In its list of the inconveniences of coitus, we again find a loss of natural
heat, a weakening of eyesight and a variety of dangers or benefits that
depend on the complexion of the person involved. The three elements
necessary (humour, heat, flatulence) are given their proper importance,
as are the post-coital discomforts noted by Constantine the African. The
psychological component, however, is dealt with more forcefully; too
long an abstinence leads to listlessness and depression in men who are
seeking to lead a holy life; on the other hand, the sexual act clarifies
thought and sobers up the man in love. The author devotes the first
part of the *Speculum* to medication, dieting and the remedies that increase
or diminish the quantity of sperm. Different enemas (one of which has
to be administered over ten successive nights of chastity) and several
ointments are recommended to improve erection. The treatise changes
tone when it comes to the nature of women. Every man must pay
particular attention to his partner's habits and to the strength of her
desire; without this knowledge, power, riches, or compliments will be of
no avail. So the author reveals the secrets of women, but his discourse
has nothing in common with the treatises that generally adopt this role.
As well as taking care to point out that female behaviour varies with
age and the period of the month, he goes into psychological details. It
is not enough for the woman to be beautiful; she must also be
knowledgeable, generous, sincere and cheerful. To discretion and cunning
she must add other qualities associated with sociability: if she is valued
by the women she frequents, this is a good sign. Then follows advice
that will enable the man to guess whether or not a woman is in love.
If her passion refuses to be disclosed, it can still be recognized by
thinness, sighing, and that listlessness which leads to anorexia and
insomnia. Here we find, applied to the female sex, the portrait of the
'heroic' lover. The woman who wants to show her passion can be
recognized by her seductive gestures, the way she bustles eagerly around

children and a vengeful fury if her lover pretends to be neglecting her. The delights of the first encounter must be carefully prepared; a meal and perfumes will help one to gain a victory, as will a declaration feigning despair. The *Speculum* diverges imperceptibly from medicine to link up with Ovid or the world of the *Thousand and One Nights*.

The author shows a pronounced taste for classification. Thus he groups into four the parts of the body that must answer to the same qualities.[117] A woman's nobility is shown by the predominance of three colours: black, scarlet and white, but also by an adequate distribution of slenderness and roundness. In the same way, a body that gives off a pleasant perfume from its four most delicate parts is a sign of female beauty. After these aesthetic considerations, the author comes to the performance of the act. Following the tradition of medical works, he here sets down various pieces of advice aimed at smoothing out the difference of rhythm between the partners. Foreplay (which the author classifies according to way of touching and erogenous zone)[118] constitute the best stimulation if the woman's desire is slow in being aroused; the only psychological mechanism is the one envisaged for the slow man, who will have to make use of his imagination. At this point in the treatise, the author tackles the problem of positions. The first considerations stray hardly at all from what was declared in Avicenna's *Canon*: divergences from the natural position are not good for one's health or for conception. But there is a new interruption in the argument and the author bluntly lists twenty-four positions that he classifies, in accordance with a constant desire to be systematic, under five principal headings. The descriptions are brief but clearly specify the way one move should follow another.[119]

The *Speculum* is the only treatise hitherto discovered which clearly describes an art of sexual positions before the Renaissance. Should this be considered an accident, as the avatar of an Arab tradition that remained occult in the medieval West? The *Speculum* differs from Eastern models derived from Taoism or from the *Kamasutra* in that it puts forward a simple manual of movements, the aim of which is no longer specified. At most, the author points out the most favourable positions for the achievement of full pleasure. But at no time is sexuality considered as a means of attaining a higher level of knowledge. The originality of the *Speculum* consists in its presenting, as early as the fourteenth century and in a Western language, a carnal technique; it thus joins a genre that later enjoyed particular favour in the Italy of the Renaissance.[120] While these descriptions made it possible to attempt to put them into practice, they gave few details on the means of reaching a real awareness of the body. What we have here is above all a classification, a sort of catalogue. Clitoral sensitivity does not appear in it. On the other hand, the author cites the breasts and the navel as among the erogenous zones. It must be pointed out that the navel is hardly ever mentioned by Western doctors, though it was referred to by Isidore of Seville.[121] It nonetheless appears as the seat of the female libido in certain treatises on physiognomy. Michael Scot, strangely enough, referred on this point to the Book of Job.[122]

Despite the licentious character that the *Speculum* took from its subject-matter, there is no trace of obscenity in it. Set firmly within a medical framework, it even appears to be particularly 'healthy'; while it mentions different positions, the natural paths are the only ones envisaged and no technique for retention is disclosed. In that, the *Speculum* differs from its Eastern counterparts and sticks to the opinions expressed in medical works of the Latin world. Although this treatise constitutes at present the sole representative of a hitherto unknown genre, one can make a case for its not being completely exceptional. Although they avoided alluding to different positions, doctors found room in their works for a great deal of advice which diverged from anatomical or physiognomical description and bore a closer resemblance to an erotic science. In it could be discerned the manifestation of a medical reaction against the dangers that threatened civilization. Epidemics of the plague increased falls in the population, and the last centuries of the Middle Ages were characterized by the rise of an apocalyptic vision; fanatical preachers foretold the end of the world. Doctors were led to denounce the danger of such influences[123] and to praise the joys of life rather than to act repressively. From the same point of view, Jacques Despars, in the fifteenth century, declared firmly that nobody should blush about their sexual life, and underlined the happiness that procreation could bring:

[The man who suffers from a diminution of coitus] should become used to not blushing about the sexual act, or about listening to stories about it. Books discussing carnal love, the acts of Venus and their figures should be read in his presence, so that he will not blush either at talking about it openly, or at hearing it discussed, or at performing coitus when he so decides. He must frequently imagine the pleasures of love and he must bear in mind that to beget one's fellow-creatures constitutes for living beings an entirely natural task; he must remember that without coitus, the human species would be extinguished, and it is pleasant to have children who will produce others and keep the world going.[124]

The conclusion of this advice is not without its biblical reminiscences, but we will emphasize the insistence with which doctors appropriated the remarks of Avicenna and recommended the reading of 'specialized' works. Perhaps the *Speculum* should be considered as the survivor of a genre represented more widely in the Middle Ages than one would be tempted to imagine on reading the theologians alone.

4

The Innocent and the Guilty

Our examination of physiological theories leads us to enquire into the place occupied by human responsibility in medical writings. The problems posed in the Christian Middle Ages by the introduction of astrological determinism are well known. The same type of question was evidently raised about medicine: accepting the existence of physiological constraints imposed limits on the exercise of judgement and on human liberty. The entire medical discourse on sexuality tended to maintain a balance between these imperatives, and to show what it was possible to demand from the human body and what it was not. While bringing out the full force behind the idea, of Platonic and Galenic inspiration, 'that the habits of the soul follow the temperament of the body', and while demonstrating, in the well-chosen formulation of the *Placides et Timéo*, that, unlike souls, 'bodies are not equal',[1] medical discourse also condemned slips in behaviour that it considered to lie entirely within the responsibility of the subject.

THE LIMITS OF FREEDOM: THE SUBJECT'S PREDISPOSITIONS

Whether they were supporters of the Aristotelian division between form and matter, or of the theory of the two seeds, the scholars of the Middle Ages considered the fact that one belongs to one sex or the other as the result of a conflict. One of the *Questions on Animals* which Albert the Great devotes to the subject gives an account of the Aristotelian option.[2] The production of a male foetus resembling the father requires 'a total victory of the male seed over female matter'. To achieve this complete masculine success, various conditions have to be fulfilled. Indeed, the sperm transmits first of all the characteristics of the species, thanks to its *virtus hominis*; it is the sperm that makes the foetus human or animal. Following the principle that every natural agent engenders as far as possible a creature like itself, male sperm tends to reproduce in another

FIGURE 4.1   *Adam, Eve and the serpent. Ms Fr. 17001, fol. 107v. (Bibliothèque Nationale, Paris)*

TABLE 4.1   *Factors determining sexual and individual characteristics according to Albertus the Great*

Child conceived	Male sperm bearing sexual characteristics	Male sperm bearing individual characteristics
Daughter resembling mother	Vanquished	Vanquished
Son with only a distant resemblance to father, but possibly resembling male ancestors	Vanquishes	Vanquished by own competent elements
Daughter resembling father	Vanquished	Vanquishes
Son resembling mother, or her female ancestors	Vanquishes	Vanquished by female matter

being the sex and the characteristics of the individual it has come from. If it is not strong enough, it may either be vanquished by the female matter and fail to transmit its sex, or else not succeed in transmitting its own characteristics, which are thus supplanted by those of the ancestors which it contains virtually. We will summarize in a table the different possibilities envisaged by Albert the Great, following Aristotle (see table 4.1).

The supporters of the two-seed theory also retain the principle of a conflict which they more often than not associate with the right–left opposition. We will quote the pseudo-Galenic *De Spermate*, for it puts forward a system close to that which will be taken up and embellished by the Salernitan authors, and by William of Conches. The range of possibilities includes the types of the virile woman and the effeminate man:

If the seed falls into the right-hand part of the womb, the child is a male. . . . However, if a weak virile seed there combines with a stronger female seed, the child, although male, will be fragile in body and mind. It may even happen that from the combination of a weak male seed and a strong female seed there is born a child having both sexes. If the seed falls into the left-hand part of the womb, what is formed is a female . . . and if the male seed prevails, the girl child created will be virile and strong, sometimes hairy. It may also happen in this case that as a result of the weakness of the female seed there is born a child provided with both sexes.[3]

Considered to be a foetus created in the middle, according to the theory that says there are seven cells in the womb, the hermaphrodite is also a child born of an indecisive conflict. It must be noted that, unlike other writers, doctors have little to say about this extreme case. They are content with quoting some anecdote, or with pointing out, as does Avicenna, that certain hermaphrodites may, as they prefer, be either a man or a woman: the Church forced them to choose. Furthermore, surgery could offer them the means of correcting the most visible anomalies. In the *Great Surgery* of Guy of Chauliac, we read:

Hermafrodicia is the nature of double kynde. And it is in men (after Albucasis) after two maneres, for sometyme it is in the place that is apperynge under the stones. In a womman forsothe there is another in the whiche a yerde [penis] and prive stones [testicles] apperen above the prive chose [vulva]. And thai ben ofte tymes cured by kyttynge, as Avicen saith, but noght that forsothe that maketh water, as Albucasis saith.[4]

Writers hardly go beyond these remarks. Sexual ambiguity lies more in the domain of natural philosophy; medicine can only testify to its own powerlessness and leave the myth to continue to nourish people's imaginations. Drawing at once on the *Metamorphoses* of Ovid and the *Achilleid* of Statius, the *Placides et Timéo* confers on the character of

FIGURE 4.2   *A hermaphrodite in the bath: the upper part of the body is double: face and bust of a woman, face and bust of a man. Contained in a manuscript of Ovid's* Metamorphoses. *Ms Fr. 137, fol. 49. (Bibliothèque Nationale, Paris)*

Hermaphrodite the power to reveal to men the secrets of women.[5]

Within the two substances, maternal and paternal, another struggle takes place, the result of which will have an equal influence on the individual's sexual life; to the innate characteristics, such as virility or an effeminate appearance, are added less visible predispositions:

Thanks to their qualities, the four humours form the sperm; this descends from the kidneys into the testicles where it is purified and completed, then it is received into the womb. Thus men and women are made of four humours. The child will resemble the humour that is superabundant in the sperm at the moment of coitus.[6]

For the author of the *De Spermate*, it is in this way that the temperament of each individual is determined. The best sexual predispositions are the prerogative of the male nature whose characteristics are the closest to the qualities required for coitus. The sanguine temperament in which warmth and humidity are dominant will quite naturally be the most appropriate. Furthermore, it brings together the qualities that preside over life, death being the consequence of a victory of cold and the culmination of dessication due to the complete loss of the *humidum radicale*. It is easy to understand that the least suitable temperament is that of the melancholic man: cold and dry, he fulfils none of the conditions required. Often subject to impotence, he is also characterized by an unbridled longing, and a twisted judgement which impels him to seek, under the impetus of violent desire, the qualities that he lacks. If he is not impotent, he will be lustful.[7] Between these two extremes are found the intermediary temperaments which include one of the preconditions that make for sexual success. Taking advantage of the opportunity to mock some of his masters, an English copyist of the thirteenth century reveals with a certain brutality, in the middle of a question of Salernitan inspiration, the predispositions inscribed in the temperament of his masters at Hereford:

- the sanguine (hot and moist) 'desire much and are capable of much';
- the choleric (hot and dry) 'desire much and are capable of little';
- the phlegmatic (cold and moist) 'desire little and are capable of much';
- the melancholic (cold and dry) 'desire little and are capable of little'.[8]

How did the people so described react? History does not say.

As a science subsidiary to medicine, physiognomy also made it possible to recognize one's characteristics in accordance with physical signs. Great medieval scholars tried their hand at this art, which they found in classical and Arabic sources, and especially in the *Secret of Secrets*. The treatise of Pietro d'Abano gives, above all, advice of a political nature; on the other hand, the *Physionomia* which Michael Scot dedicated to the emperor Frederick II devotes a good deal of space to sexual life. In conformity with the Arab tradition which puts forward advice designed to help a man choose women in the harem, Michael Scot draws female portraits.[9] The same criteria can naturally not be applied to both sexes; but, as in man, the highest degree of heat must be sought in woman. The best predispositions are found together in the young girl of more than twelve years who has lost her virginity. Small and firm breasts, thick hair in the right places and a highly coloured complexion, are good signs. Such a woman likes to behave insolently, shows no signs of piety and is capable of getting drunk; she enjoys singing, going for walks and having fun. She is in a permanent state of desire which she can satisfy in the sexual act. Since her menstrual blood is not very copious, her periods are irregular and she rarely becomes pregnant. Michael Scot adopts the theory of the double seed; according to him, the less menstrual blood a woman produces, the more abundant a supply of sperm she has and the more pleasure she feels. As a consequence, she produces only a little milk, which explains the small size and the firmness of her breasts. At the opposite extreme there is the woman whose nature is colder; often too young, her breasts are big and soft, her hair rather sparse and her complexion pale. Timidity and reserve are the qualities that characterize her, as well as piety and credulity. She only rarely manages to satisfy her desire; her periods are regular, she becomes pregnant easily and has abundant milk. Behind these two portraits sketched by Michael Scot one can make out the profiles of the courtesan and the mother – the former of whom was, we may suppose, more greatly valued by the entourage of Frederick II. But the remarks relating to menstruation give reality and physiological justification to these two extreme female types. We again come across the aesthetic importance granted to women with firm, small breasts; Nicolette, in this respect, met all the conditions of medieval beauty:

She had fair and tightly curling hair, laughing grey eyes, a straight nose set in an oval face, and red lips, brighter than any cherry or rose in June; her teeth were small and white, and her small, firm breasts lifted the bodice of her dress as they might have been two walnuts, while her waist was so tiny that you could have ringed it with your two hands.[10]

The originality of Michael Scot's text in comparison with the Western tradition as a whole is that it links the aesthetic type of the small-breasted woman to a better aptitude for sexual life.

Over and above the characteristics linked with temperament, the qualities of which can also be recognized by touch, physiognomy helps one to surmise the aptitudes of the hidden organs by observing the visible signs. The *Physiognomies* of Greek and Arab origin fix, like a stereotype, the portrait of the lustful man. Pseudo-Aristotle ascribes to him white skin, straight, thick, black hair that falls right across the temples, and staring eyes.[11] In the second book of the *al-Mansuri*, Rhazes paints the same picture, adding a reddish complexion that alternates with pallor.[12] While the eyes perform the classic function of acting as the messengers of the soul, other parts of the body are deemed to transmit information, according to a more 'scabrous' tradition. Michael Scot and even Michael Savonarola, in a perfectly serious *Mirror of Physiognomy*, linger with some pleasure over the significant size of the nose in the man and the foot in the woman.[13] We find in the work of Michael

Savonarola, the traditional pun on Ovid, nicknamed 'Naso', being given an important place. Let us add that ever since Pseudo-Aristotle, a snub nose was considered to be the prerogative of lustful men.

We could go on multiplying the external signs supposed to reveal sexual capacities; the description of the male and female organs considered to be the most suitable also occupies a far from negligible place in these works. It is enough to remember that the Middle Ages created a perfect synthesis between physiognomical data and the theory of temperaments; the work of Michael Savonarola represents, from this point of view, a culmination.[14] As a privileged opportunity to tackle the theme of the relationships between body and soul, the physiognomic discourse which, as Michael Scot recalls, must remain secret because of its effectiveness, claims also to transmit a sexual initiation in which the expert in medical knowledge poses as master.

The constraints imposed by the individual's temperament are not absolute. Medicine is capable of rectifying its excesses and of modifying its imbalance in order to obtain a precise effect: this is the very foundation of medieval therapeutics, based as it is on the pathology of the humours. By increasing or diminishing the quantity of one of the four humours, one acts on the temperament of the whole body or of a definite part of it. But since this effect is only momentary, it may even be decided to intervene before the conception of the future embryo. Even the possibility of intervening before conception, in order to choose the child's sex, is foreseen – a theme that these days is once more occupying an important place. From the same angle Albert the Great raises the following question: 'Is it possible for a male or a female to be created artificially?'[15] The answer is positive: one can act before copulation, since sexual differentiation is determined merely by accident. The chances of obtaining a son will be increased if the virile heat is reinforced by an appropriate medicine. Another kind of action consists of favouring the right-hand side in the emission or reception of the semen.

Other factors must however be taken into account – factors which, linked to external influences, have a bearing not only on the way sex is determined, but on one's love-life as a whole. We will leave aside the problem of astrology which lies outside the limits of our study, and discuss instead the modifications brought about by the climate. In line with a tradition that goes back to Hippocrates, Arab authors lay a particular stress on the importance of meteorology. Desire and sexual potency vary with geographical location and the rhythm of the seasons. Medieval doctors adopt without any major changes what they read in the *Pantegni*, and then in the *Canon* of Avicenna. In accordance with the rule *similia similibus*, the best season for the pleasures of Venus is the spring, since it is hot and moist in nature, and is thus the time when the sanguine humour predominates. Summer and autumn, during which bile and melancholy abound respectively, are hardly propitous times: performing coitus in these seasons leads to excessive dessication. Winter is less unfavourable: close in nature to phlegm, its cold quality is compensated for by an abundance of moistness. These principles, drawn from a law that brings similar things together, must however be somewhat modified by another principle which, based on physics, states that opposites attract. The inhabitants of cold regions are braver than others and, endowed with a greater internal warmth, they are full of 'digested' sperm. Doctors explain this fact by referring to the ancient concept of *antiperistasis*, which is still used as an illustration by numerous medieval authors as a means of accounting for the movements of projectiles.[16] As used in physics, the principle explains why one of the elements coming into contact undergoes a sudden increase in its dominant quality. The same process takes place in breathing: a shock occurs between the cold air inhaled and the internal heat, which is compacted and becomes more intense. This also explains the different ways the two sexes behave. Ever since Aristotle, doctors asserted that the man has a greater sexual capacity in winter, whereas the woman is more ardent in summer. This belief risks coming into contradiction with the principles of the law *similia similibus* that is applied to season influences. So the distinction has to be drawn between determinism on the species level and the characteristics proper to each sex. The difference in nature between men and women subjects them to a discordant biological rhythm. Because of its temperate biological nature, spring may bring them together; autumn separates them irremediably. For the other seasons, Albert the Great finds a solution which allows a balance to be established: in winter, it is the man who takes the initiative in sexual pleasure, in summer it is the woman.[17] So a compromise is possible between these two beings that are so different in nature.

Guilty Imaginings

If there is one area in which doctors come up with reasons both for excusing and for repressing, it is that of solitary pleasure. Their position remains ambiguous as far as the moral responsibility of the subject is concerned. It can also be seen that they do not tackle male and female weaknesses with the same degree either of frankness or of precision. Revealingly, the authors shift their standpoint depending on the sex they are discussing. Despite the attention paid to the loss of female seed, it is evident that male self-abuse is treated at greater length. What is less easy to justify from a scientific point of view is the discrepancy between a relative silence on the reality of the practice of male masturbation and an eagerness to give information about female habits. Solitary pleasure is one of the themes on which the most misogynistic authors can give free rein to their indiscretion.

The fact that sexual predispositions are laid down in advance in accordance with temperament shows that certain individuals will encounter greater difficulties in their quest for chastity. To the general tendencies of the organism as a whole can be added the characteristics proper to the genital organs; these features show discernible variations from one individual to the next. Those least predisposed to continence are men, whose organs have either a hot and dry complexion, or a hot and moist complexion. This is clearly stated in the *Techne* of Galen and in the *Canon* of Avicenna.[18] The predominance of hot and dry qualities predisposes one to desire in all its force: the individuals thus characterized feel a continual need to copulate and they are moved by a violent desire, a *furia*. The excess of hot and moist does not have the same consequences, but it makes total continence just as harmful. Because in this case the sperm ducts are plethoric, an active sexual life does not entail any damage, whereas prolonged abstinence brings grave problems in its wake. The risks run by individuals whose organs are dominated by the complexions that we have just referred to are, respectively, satyriasis and gonorrhoea.

Medieval authors add little to the ancient picture of the first of these disorders. They repeat the classical distinction between satyriasis and priapism.[19] Both give rise to a permanent erection, but the one is accompanied by desire, the other not. As opposed to priapism, satyriasis is assuaged by performing the sexual act. Its causes are many and varied: an excess of wind due to the absorption of flatulent food, the habit of sleeping on one's back and prolonged abstinence. The *Breviarium Practice*, whose interest lies in the fact that it was written in a monastery, gives us an eyewitness account of monastic life at the end of the thirteenth century:

In different monasteries and religious places one comes across numerous men who, sworn to chastity, are often tempted by the Devil; the principle cause is that every day they eat food that leads to flatulence. This increases their desire for coitus and stiffens their member: that is why this passion is called satyriasis. . . . As I am writing this book in the Cistercian monastery of Casanova, for the love of the abbot, the prior and the monks, I will say what may be of use *ad removendum coitum*.[20]

Following the medical and monastic tradition, the author proposes a diet aimed at removing flatulence and diminishing the quantity of sperm, but also recommends simples with calming virtues such as lettuce or poppies. Baths will chill the organs responsible, as will the habit of sitting on a cold stone. The author takes from Avicenna the advice to walk barefoot: Gentile da Foligno saw the same reason behind this procedure being adopted by the Franciscans.[21] Although ancient medicine mentioned the possibility of female satyriasis, the writers of the Middle Ages gave little credence to it. At the beginning of the fifteenth century, however, Valescus of Tarentum took up the idea, asserting that under the sway of an 'unbridled appetite', women could suffer a hardening of the vagina which made that organ 'sinewy, gristly and extremely sensitive'.[22]

If few details were added to the picture of satyriasis and priapism in the Middle Ages, gonorrhoea was a central topic of lengthy discussions. This must clearly be related to the abundant literature that theologians devoted to the problem of self-abuse. Indeed, few and far between were the doctors who did not find room in their works for a chapter on 'gonorrhoea', a term that covers more the involuntary loss of sperm – spermatorrhoea – than blennorrhagia.[23] This chapter was more often than not followed by another on 'nocturnal emissions'. As we have seen in regard to female leucorrhoea, doctors did not manage to establish any hard and fast distinction between functional or pathological disorders, nervous problems and deliberate self-arousal. Certain of their remarks could meet with indulgence on the part of a theological reader, but others, through the importance they granted to the imagination, drew down repression on themselves.

The causes of gonorrhoea were many and varied. The most compelling consisted in the predominance of a warm and moist complexion; the doctor, who in this case advised marriage, implied that no other remedy could be effective.[24] Gonorrhoea could also be the result of a spasm, as in the epileptic. But most of the time, the aetiology presented a less serious character: an appropriate diet would diminish the quantity of sperm, as would blood-letting. In the *De Consideratione Operis Medicine*,[26] Arnald of Villanova insisted on the necessary harmony that the doctor must maintain between these two therapeutic procedures. To prove his case, he went into the detailed description of a particular instance. A cleric was suffering from chronic emissions of sperm because of his hot and moist complexion. The monastery doctor could not manage to cure him of this complaint. For after each bleeding, the monk, far from being relieved, fell prey to a resurgence of the malady. An interrogation revealed that his diet was wrong. Ordinarily, the monk took his meals in the refectory: he ate beans, cabbage and dry cheese, substances that increased semen. After being bled, it was in the infirmary of the monastery that he took his meals; here, he was given food meant to enable an individual recently deprived of part of his blood to recoup his forces – eggs, wine, meat stock and fat cheese. Over and above the fact that food such as this also created sperm, its action was doubled or quadrupled by the fact that after bleeding, the testicles aspired to attract a large amount of matter transformable into semen. The doctor's error consisted in treating the monk, after surgery, in accordance with the general rules that determined the diet of a recently bled patient, and not in accordance with the demands of his complexion and his illness. After the phlebotomy the patient should have been prescribed foodstuffs to cool him down, such as lentils in vinegar, lettuce or salt fish. This evidence on a real case from monastic life shows the difficulty, indeed the impossibility, of imposing chastity on men who were not naturally predisposed to it. Writers stated repeatedly that continence was dangerous for young people and for men endowed with a hot and moist complexion. The matter not expelled affected the heart: it gave rise first of all to anxiety and depression, then, as a consequence, it damaged all the other functions.[26] We notice, however, that while doctors insisted on the dangers of an excessive retention of semen, they did not explicitly recommend that men have recourse to masturbation.[27]

The observation of monastic conditions weighed heavily on medical discourse. Even if they stuck to the ancient definition of gonorrhoea, that is, an involuntary emission without concupiscence, authors no longer made this distinction in the discussions they devoted to the subject. The ambiguity was long standing and may be linked to the advice of certain theologians, such as Vincent of Spain or Anthony of Florence, who recommended that the doctor should not reveal to monks the cause of their illness, but that they should, rather, look after their diet.[28] The importance granted to the religious problem of emissions was an obstacle to medical thinking about pathological discharges. Such confusion overwhelms the descriptions that the historian of medicine cannot detect any potential traces of blennorrhagia. Only a few remarks lead us to suppose behind these 'gonorrhoeas' a more serious infection. Thus, Bernard of Gordon reveals that many men die because they do not dare to reveal such a shameful malady and Valescus of Tarentum quotes the case of diurnal emissions occurring with alarming frequency.[29] Despite these rare signs, the nosological reality becomes blurred when faced with the demands of the religious and social context. The doctor should help his patients to lead a chaste life, by showing them what lies within the responsibility of the body.

The definition of coitus given by Gerard of Bourges may also illustrate the medical position on the problem of self-abuse: 'Note that four things play a part in coitus: a powerful imagination, a warm and dry complexion which stimulates the member, a flatulence that gives rise to an erection, and spermatic matter. If one of these things is missing, there is no real coitus.'[30]

It is the presence of these four elements that distinguishes self-abuse from purely physiological gonorrhoea. The conclusion reached by doctors seems clear and coincides with that of the theologians: for there to be responsibility on the part of the subject, the emission must be brought about by the power of the imagination. Let us recall what Albert the Great wrote on the subject of the loss of virginity.[31] Simple natural purgation does not corrupt the body and neither does mere imagination, if it is not followed by an emission; on the other hand, virginity is lost if an involuntary emission occurs after thoughts about coitus or the pudenda. So the doctors consider a wandering imagination to be among the external causes of gonorrhoea. In this case, dieting or letting oneself be bled are of no use; one must undergo fasting and flagellation. The author of the *Breviarium Practice* takes an even firmer stand on the side of a therapy of the soul, recommending, as do the churchmen, a 'sublimation' of desire: the best way of avoiding self-abuse resulting from imagination is to contemplate the first cause and concentrate with all one's might on the love of God.[32] One of the men who experienced this transference from one sort of love to another in the Middle Ages was Raymond Lull; having been 'crazy with carnal love', he became, at the age of around thirty, 'crazy with spiritual love'. The experience gave support to his medical and philosophical learning:

When one's thoughts start to imagine the beauty of women, lust extends and spreads throughout the sensitive power: and the more frequently it imagines, the more strongly the sensitive soul is filled with a burning lust. Thus, the man who wishes to extirpate and annihilate lust from his sensibility should imagine other things which are neither beautiful nor attractive.[33]

This way of diverting thought was prescribed in ancient medicine. Caelius Aurelianus discussed the problem of nocturnal emissions which accompany erotic dreams in these terms:

Nocturnal emission is the discharge of semen in the course of a dream. The name [*oneirogmos*] is derived from this symptom, for the dream [*oneiron*] provides the venereal stimulation which gives rise to the emission. But, in general, nocturnal emission at the outset is not a disease or even a concomitant (Greek *symptoma*) of disease, it is essentially a consequence of what the person sees (Greek *phantasia*) while he sleeps, and results from a longing for sexual enjoyment, that is to say, from constant and uninterrupted sexual desire, or, on the other hand, from continence or a long interruption of sexual activity.[34]

Even while asleep, the subject is still responsible since 'The patient's sensations in waking life readily give rise to dream images in which the movements resemble actuality.' Caelius Aurelianus imagined, in addition to the treatment of the body, a cure of the soul which consisted principally in diverting the mind from its preoccupation with sex.[35] With regard to nocturnal emissions, writers in the Middle Ages merely subscribed to this medical tradition – all the more so as it was in perfect agreement with the demands of Christianity. Nevertheless, the line of demarcation between emissions considered natural and those considered depraved remained uncertain. The fixing of images in the *phantasia* could result from deliberately dwelling on a particular thought, but also from the chance sight of an attractive person or of a couple of lovers, without there being any voluntary thinking on the part of the subject. It could happen that the imaginative virtue moved the expulsive virtue and caused a spermatorrhoea, just as a harsh noise sets the teeth on edge, or the sight of a person vomiting or urinating makes one want to do the same. Here we are in the domain of sensation, not of the will;[36] the hybrid status of the imagination, the sole mental faculty mentioned by the authors, left some doubt as to the responsibiity of the subject, even if the cure recommended was extremely severe and even if in this case the doctor considered the emission to be a sin which sullied the soul at the same time as it did the body. A lack of precision was obligatory in a domain in which the evaluation of responsibility was incumbent on the individual alone. We will quote Michel Foucault discussing the Twenty-Second Lecture of Cassian:

Self-abuse was not simply the object of a stricter ban than the others, or more difficult to observe. It was a measure of concupiscence, in so far as it was possible to determine, throughout the activities that made emission possible, prepared for it, encouraged it and finally unleashed it, what, among the images, perceptions and memories in the soul, was the proportion of voluntary and involuntary factors. The monk's efforts to keep himself under control all consisted in never allowing his will to become involved in that movement from body to soul and soul to body over which his will might have some influence, encouraging it or arresting it through the power of thought.[37]

The fragility of the medical arguments could nonetheless open the way to a kind of casuistry, bringing directly into play the way one controlled one's intentions. At the end of the fifteenth century, the theologian John of Wesel came into conflict with the ecclesiastical authorities by discussing the problem in terms such as these. In the advice he addressed to a Carthusian on the purgation of the kidneys – organs considered to be the seat of the male libido – he went as far as to state that the expelling of semen was not a sin if it came about by physical necessity and if the will was not directed towards concupiscence.[38] Many people, he said, make a vow of chastity without a real awareness of their own strength and fall ill as a result of abstinence. Now an illness born of continence is cured only by its opposite. Thus, if the emission is brought about for health reasons, the soul remains chaste. When incontinence, even deliberately brought about, remains strictly bodily, the spiritual value of continence is not besmirched. Medical thinking did not go as far as that. While showing the dangers of excessive abstinence for certain temperaments, and while recommending a diet that reduced the quantity of male semen, it refrained from envisaging the deliberate expulsion of that semen.

On the other hand, reference to solitary practices was made in the advice meant for women. The most frequent remarks were those accompanying the treatment of hysterical illness. We will be returning to this point. But the same type of comment was made on other occasions, for example during a discussion of the young girl who has recently attained puberty. The most precise treatment of this topic is to be found in Albert the Great:

Around the age of fourteen, because of the menstrual blood and the spermatic humour descending, the thighs start to thicken, the slit in the vulva closes, the labia grow softer and thicker, and a soft down springs all around: these are the signs of puberty. The girl then starts to desire coitus, but in her desire she does not emit and, the more she performs the sexual act or has recourse to manual practices, the more does she desire; so much so that she attracts the humour, but without emitting it. With the humour, heat is attracted and, as the woman's body is cold and her pores closed, she does not emit the seed of coitus rapidly. That is why certain girls around the age of fourteen years cannot be satisfied by coitus, and if they have no partner, they imagine coitus or the male member, or indulge in practices involving the use of their fingers or other instruments until their channels are opened by the heat of the friction and then the spermatic humour comes out, together with the heat that accompanies it. Thus their groins are tempered and they become more chaste.[39]

It would be difficult to put things more clearly: in the young girl, masturbation is often a physiological necessity. It even helps one to acquire chastity. In the older woman it has its uses too. According to Avicenna, it allows the rhythm of a slow woman to be harmonized with that of a more rapid man.[40] While Western commentators hardly wasted much time in justifying the usefulness of these caresses, they had no hesitation in describing the way they were carried out. Numerous writers found in the therapeutic treatment of hysteria or sterility an opportunity for alluding in precise terms to the instruments – the substitutes for the male member, made out of different kinds of material. One of the most explicit writers is probably the author of the *Breviarum Practice* who, you will remember, confesses that he is writing his work in a monastery. He reveals the 'sodomite' practices of the wives of Italian merchants, and claims that they use a range of appropriate accessories.[41] A social reason for this is given; the husbands undertake long journeys on business and their wives dare not commit adultery for fear of becoming pregnant. It is noteworthy that the same author enumerates, on the border between burlesque comedy and sadism, the subterfuges to which young Neapolitan wives had recourse so as to feign virginity on their wedding nights; among them one finds, for example, the prior application of leeches.[42] He is not the only one to show a certain enjoyment in discussing such matters: masturbation and ways of quickly patching up virginity constituted an opportunity for a sexually repressed cleric to refer with precision to the female organs, while at the same time reaffirming the perversity of the other sex. In addition, apart from the occasional allusion by Albert the Great,[43] the descriptions of female solitary practices hardly seem to leave any room for clitoral sensitivity; thanks to the explicit mention of objects, reference to masturbation is such as to cast no doubt on the supremacy of the male organ, for which all the rest is merely a half-effective substitute.

While medicine attempts to conserve male semen and reduce its production in order to avoid any loss, it encourages the expulsion of female liquid, retention of which can be extremely harmful. Over and above the suffocation of the womb and sterility, it leads to other disorders, such as a swelling of the stomach or the formation of a mole.[44] So male and female involuntary emissions do not have the same seriousness. The woman, because of her cold nature, cannot engender a noble substance; humour is transformed into venom if it is retained, and it is better to expel it, even by means of a sinful procedure. As for the sin of Onan – the loss of semen in the man – it constitutes a sin not only from an individual point of view, but from the point of view of the species. Bernard of Gordon recalls that 'gonorrhoea' leads to the dying out of humankind, since it is 'like the stream of humanity'.[45] The manuscripts show on this point an extremely revealing confusion in spelling: in numerous works, including the *Viaticum* of Constantine the African, *gonorrhea* becomes *gomorrhea*.[46] This confusion reaches its logical conclusion in Valescus of Tarentum: 'Gomorrhoea is taken from the word Gomorrha, because of the vain sheddings of human semen that occurred in that city'.[47] Every non-fecund emission is an irreparable loss for humankind.

Even when, from a Galenic point of view, a generative virtue is ascribed to the female seed, there is no equality between the two seeds. Referring to the sixteenth and seventeenth centuries, Jean-Louis Flandrin notes one of the consequences of this inequality: while for most theologians, female retention after male ejaculation constitutes a sin, man on the other hand is not required to wait until his partner emits her seed.[48] As for Arab thought, it managed to avoid giving more importance to the male semen. As al-Ghazalî puts it:

The two sorts of sperm thus behave, respectively, like offer and acceptance in the judicial existence of contracts: the one who makes an offer which is withdrawn before it is accepted, does not commit any breach of contract in annulling it or cancelling it; on the other hand, every time offer and acceptance both take place, the act of withdrawing the offer is indeed a breach of contract, a dissolution and annulment (both illegal) of it. The child cannot be formed from the male seed while this is still in the man's vertebrae; the same is still true at the moment this seed is expelled, so long as it has not mingled with the sperm of the woman or with her (menstrual) blood.[49]

Thus is raised, in very clear terms, the question that met with no logical answer in the thinking of theologians and doctors. This judicial comparison occurs in the discussion of al-Ghazalî on how to limit births. He opposes certain doctors of the law and takes up the defence of *coitus interruptus*. The precise moment at which life begins is that of the union of two seeds; any male or female emission cannot be reprehensible so long as there has been no conjunction of the two elements. Sexual liberty and the respective role of the partners are safeguarded, as is, at the same time, respect for the initial state of the embryo. On the basis of Aristotelian thought, theologians destroy this balance and the most liberal of them are often reduced to mongrel solutions involving strange omissions. In a system of thought that does not accept contraception, but tolerates prostitution, there exists a sexuality of which nobody speaks, a sexuality without consequences. The prostitute is sterile, for evident physiological reasons, but that settles things well in theory. She it is who is responsible for the wastage of male semen, because her womb is not capable of retention. In the Middle Ages, there were in fact two systems of representation of sexuality. Outside marriage, certain forms of behaviour apart from sodomy were tolerated or practically ignored. On the other hand, as soon as man and woman came together as a couple, the person of the woman was alienated and subjected to waiting for the male semen. It is clearly easy, in the majority of cases, to draw a parallel between her dependency, her absence of any erotic existence and her situation within society.

The ideas of Galen or Avicenna were not enough to reverse this tendency. The absence of a recognition of the cyclical nature of female fertility, and the link between fertility and a disordered and irregular secretion, contributed to maintaining the imbalance between the two types of seeds. From a natural point of view, female masturbation could be envisaged as an individual necessity without any consequence for the destiny of the species. It was acceptable from a social point of view. Numerous women went through long periods of solitude, while their husbands were off at war, on pilgrimage, or engaged in commercial transactions. The implicit acceptance of female masturbation enabled the purity of the lineage to be safeguarded.

The Excluded

Even more than in the case of solitary practices, doctors showed a great reticence in tackling the problem of sexual relations that diverged from the religious and moral norm. In this caution we can see first and foremost the expression of a certain self-censorship, but we may also suppose that, since scholars gave no physiological, or even pathological, explanation for such behaviour, it was because they laid the entire responsibility at the door of the subject. Certain deviations lay outside the realm of medical scrutiny and remained within the domain of ethics.

We can find an example of deliberate omission at the end of the eleventh century in a discussion of homosexuality. During that period, the *Aphorisms* of the Nestorian doctor Yaḥyâ ibn Māsawaih (John Mesue) were translated, probably in Italy. The Arabic text sets out two propositions:

The man who often penetrates from behind risks having born to him from these women sons who are effeminate or dainty, especially if the woman requests it.

The man who experiences no pleasure with women and enjoys more often than not congress with blameworthy men should suspect that there is something wrong with him.[50]

These two aphorisms, despite the fact that they express a rather critical attitude, are the only two which the translator omitted. This is clearly no coincidence.[51]

To understand the intellectual position of the doctors, it is a good idea to recall the theory of unnatural pleasures set out by Thomas Aquinas in the *Summa Theologica*. Among the pleasures considered to be unnatural, a distinction must be made. Some are 'connatural' in certain respects; following the corruption of the natural principle in a particular individual, something that is contrary to the nature of the species becomes 'accidentally natural to this individual'. The corruption may come from the body; thus to somebody suffering from fever, sweet things seem bitter, and somebody who has an evil temperament will take pleasure in eating earth or coal. The corruption may also come from the soul: 'Sicut propter consuetudinem aliqui delectantur in comedendo homines, vel in coitu bestiarum aut masculorum, aut aliorum huiusmodi, quae non sunt secundum humanam naturam.'[52]

This distinction clearly establishes the limits of medical competence. The corruption of taste and judgement entailed by illness or a fault in temperament fall within its scope; the eating of earth or coal referred to by Saint Thomas appear in the medical texts of the Middle Ages: moreover, modern medicine recognizes in it a symptom of metabolic deficiency. But homosexuality, placed in dangerous proximity to cannibalism and bestiality, was considered neither as an illness nor as a defect of temperament. The statement of predispositions following the conditions of conception did indeed find room for intermediate sexual types, such as the effeminate man and the virago, considered to be sterile, but these characteristics did not necessarily imply homosexual behaviour. This latter came from a corruption in the soul and doctors found no physical basis for it.

Yet homosexuality was not totally absent from medical literature. The *Canon* of Avicenna referred to it on several occasions and Gerard of Cremona did not think it right to censor the passages in which it appeared. Avicenna followed the Arab tradition: he expressed no criticism; he merely compared homosexual and heterosexual relations, and described certain behaviour that he considered tainted with perversity. How were these remarks perceived by Western doctors? It is not easy to tell, since writers did not generally take up these comments in their works: they doubtless formed part of the 'illicit and indecent' remarks that Bernard of Gordon attributed to Avicenna. So we must examine the commentators on the *Canon*.

We have already had an opportunity to point out that Gentile da Foligno, at the beginning of the fourteenth century, avoided commenting too explicitly on the passages that appeared scabrous to him. He does not repudiate this tendency as far as homosexuality is concerned and skirts admirably round the question. Thus, when Avicenna compares the respective advantages and disadvantages of relations with women on the one hand and paedophilia on the other, the Italian scholar contents himself with stating: 'Here Avicenna is comparing different forms of coitus. *Quam vulva*, that means *coitus in vulva*.'[53] No information filters through this commentary. Another passage in the original could not completely be evaded, because it takes up a complete paragraph; in it, Avicenna discusses passive homosexuality.[54] Through a regard for decency or else through ignorance, Gerard of Cremona simply transliterates the name of this 'illness' which appears, depending on the

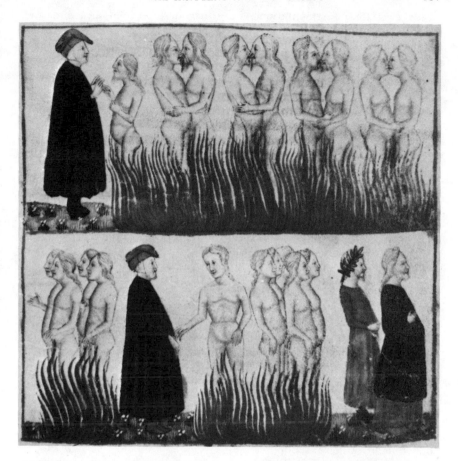

FIGURE 4.3   *The lustful. The penitents of natural and unnatural lust. From* The
Divine Comedy. *(Purg.* XXVI). *Ms Biblioteca Nazionale Maiciana di Venezia Hal.
ix. 276. fol. 45r (Photo. Foto Toso)*

manuscripts, in the form *alubuati* or *aluminati* to translate the Arabic *al-
liwaṭ* (sodomy). The translation of Avicenna's discussion of it leaves no
room for ambiguity, however: this problem affects men who, endowed
with a strong desire, but impotent when it comes to ejaculation, enjoy
contact with other men. Gentile da Foligno demonstrates a rare virtuosity
in managing to comment on this text without mentioning what it is
about.[55] He summarizes the meaning of the paragraph in these terms:
'Here the author is talking about a certain accident which happens to
men during coitus.' Gentile da Foligno then goes into a word-by-word
explanation of the passage, in the form of simple correspondences, which

leads to a total fragmentation of the discourse without any semantic reorganization. The only remark that is granted an explanatory sentence is the following: '*Naturale*, that is: that their bad thoughts have to be corrected'. Mental perversity, which can be found mentioned in Avicenna too, constitutes the sole characteristic worth explaining. Gentile da Foligno shows that he has understood, but does not want others to understand, by referring, without quoting its title, to problem IV, 26, of Pseudo-Aristotle: 'Why do some men enjoy sexual intercourse when they play an active part and some when they do not?'[56]

On *alubuati*, the commentary of Jacques Despars is more explicit.[57] Right from the start, the fifteenth-century doctor gives us evidence taken from his personal experience: when he was a *submonitor* in a secondary school, there was a cleric who indulged in such practices with the schoolboys. According to the commentary, retribution was swift: the future doctor warned the headmaster of the school so that 'such detestable and vile sodomite activity should not take place under his very roof'. After this anecdote, Jacques Despars gives a very faithful commentary on Avicenna's text, taking care to add adjectives such as 'shameful' or 'indecent'. Avicenna points out that certain scholars think that men afflicted by this form of impotence suffer from an anomaly in the nervous system of their penis; the Arab doctor considers on the contrary that the cause is not natural, but 'meditative'. He also cites arguments that are grist to the Thomist mill, asserting that 'Those who want to cure such defects are stupid.' The only possible treatment consists in destroying desire by applying physical and moral brutality: depression, hunger, insomnia, isolation and flagellation. Jacques Despars is quite naturally inclined to expand on this theme and to add a few details on the punishment of these 'sodomites':

The first means is the depression that one must harshly rouse in them through invective, criticism, and hatred of their thoughts and their vile acts. The second means is torturing them with a strong and constant hunger. The third is tiring them greatly by means of depriving them of sleep. The fourth is throwing them into a horrible prison. The fifth is beating them frequently, until the blood flows, with whip or rod.

We note, then, that the two Western commentators show a completely different attitude: Gentile da Foligno avoids explaining too clearly, but he refers to the *Problems* of Aristotle, which give a scientific justification for sexual passivity; Jacques Despars takes up with great precision all the themes alluded to by Avicenna, but he strengthens the condemnation pronounced by the latter. The same attitude can be detected with regard to paedophilia, a subject that we saw Gentile da Foligno evading completely. Avicenna's discussion is in any case not without a certain ambiguity: he states first of all that such a habit is considered by many people to be shameful and contrary to the law, then he compares it, without taking into account what he has just said, to heterosexual relations. Intercourse with young boys is more harmful in that it demands an intense effort, but it is less harmful in that it entails a weaker emission. In his commentary, Jacques Despars clearly lays great stress on condemning every homosexual relation;[58] he recalls the inevitable example of Sodom and compares homosexual practices with bestiality. Nonetheless, he faithfully records the advantages and disadvantages mentioned by Avicenna and contents himself with adding several qualifying adjectives such as 'shameful' or 'detestable', while at the same time praising female beauty, which offers greater attractions. He concludes his commentary on the passage in these terms:

It would be possible at this point to relate several types of sodomite coitus, which men and women abusively indulge in, and establish between them comparisons on their advantages and disadvantages, but I judge it better to keep silence, so that human nature, inclined towards evil and towards the exercise of new lusts, may not attempt, on hearing about them, to put them into practice and thus prejudice one's honour and one's soul.

Jacques Despars clearly raises the problem of information: by being too explicit, a warning becomes an encouragement. Confessors encountered the same difficulty, and in Bartholomew of Exeter we find a similar formulation: 'We have heard of men and women who, having heard crimes unknown to them being named, fell into sins which they had hitherto been unaware existed.' Confessors' manuals suggested prudent ways of talking.[59] On the question of homosexuality, most doctors in the Middle Ages chose to remain silent. Jacques Despars, despite the caution he proclaimed, chose to inform by explaining Avicenna's text; in the fifteenth century, the doctor's social role in fact became more extended and certain questions could not remain unanswered.

We again encounter this need to inform with regard to female homosexual relations which Avicenna predicts will occur when the man has not been able to give satisfaction.[60] In such a case, no emission on the woman's part has taken place and conception cannot supervene. On this point, Jacques Despars clearly analyses Avicenna's text, but his conclusion reaffirms the doctor's duty;[61] he must give advice so that relations beween men and women may be improved, and recourse to 'illicit' practices avoided. We again encounter the possibility that the doctor may transmit erotic knowledge, but we also see that the condemnation of female homosexual practices is not particularly severe. The adjectives 'shameful', 'detestable', 'vile' are not used and, after the description of certain practices, the commentator contents himself with adding that there exist 'other procedures which it is better to keep quiet about than to disclose'. This relative tolerance forces us to make the same observation as for masturbation: the loss of female seed has less importance for the preservation of the species. Medical literature is not alone in suggesting this kind of conclusion. Jacques le Goff quotes a rather revealing example; a young woman in Purgatory gives the reason for the pains she has to endure:

Until today I was in the grip of no slight punishment, for when I was still of tender age I gave in to indecent lust and committed shameful acts with girls of my age, and alas! having forgotten them, even though I confessed to a priest, I did not submit to the judgment [of penance].[62]

She is in Purgatory, and that is not a hopeless position. The new structure taken by punishment allows this female sinner to escape damnation. It is true that the fault was committed at a 'tender age'. Another type of evidence confirms one's impression that greater indulgence was shown towards women. Here, from Jean-Louis Flandrin, is the division of sins given by the list of reserved cases in the diocese of Cambrai, for the years 1300–10:

Reserved to the Bishop: sins against nature in a man aged over twenty.
Reserved to the penitencers: sins against nature perpetrated by women of every age and by men aged under twenty; manual self-abuse at every age.
Under the jurisdiction of the parish priest: unbridled copulation with women; the sin of self-indulgence; sins against nature in childhood, in boys up to the age of fourteen and in girls up to the age of twenty-five.[63]

So the age of 'childhood' extends, in the case of woman, up to the age of twenty-five. In addition, for her, sins against nature never seem to be reserved to the Bishop. From this, Jean-Louis Flandrin concludes: '[Homosexual relations] are less serious in girls than in boys, and that seems to me to be in conformity with what is known about judicial practice'. It also seems to conform to the role ascribed to the woman in conception, and to her submission within the family and society. If she is not a nun, the woman scarcely has any possibility of choosing celibacy and, once she is married, she will not be able to evade her obligation as a wife. Recourse to practices 'against nature' will only be occasional and will put neither the future of the species nor civilization in peril.

    John Boswell summarizes the position of Albert the Great on homosexuality.[64] Habitually inclined to indulgence as far as sexuality is concerned, the Dominican shows the greatest firmness when it comes to sodomites: 'It must be said that the deformity of sins is often measured in accordance with three [criteria], namely, grace, reason, and nature.

The sin which is against grace, reason, and nature is the greatest: that is the case with sodomy.'

Albert the Great defines elsewhere what is covered by the often ambiguous term of 'sodomy': 'A sin against nature, of a man with a man, or of a woman with a woman'. So there is no doubt about what is being condemned. No 'natural' explanation is provided by Albert. Precise biblical references are what lead him to state that the malady is innate and reputedly contagious:[65] 'The wicked are estranged from the womb: They go astray as soon as they be born, speaking lies' (*Psalms*, LVIII, 3) and the injunction given Lot by the Angels for him to leave Sodom: 'Escape for thy life; look not behind thee, neither stay thou in all the plain; escape to the mountain, lest thou be consumed' (*Genesis* XIX, 17). Homosexuality lies within the domain of sin, of a foetidness that can pass from one being to another and which has a tendency to strike high: it affects highly placed people more often than the humble.

The world of animals should have enabled Albert the Great to ask whether there might not be a natural origin to homosexuality. On the contrary, he refrains from granting any reality to the morals traditionally attributed to the weasel, the hare and the hyena. On the topic of this latter, John Boswell picks out an involuntary contradiction in the *De Animalibus*. Although he disproves the animal's bisexual character, Albert gives a magical recipe in which the fur of the *alzabo* plays a part. But he has not recognized the hyena behind this transliteration of the Arabic *al-ḍab'*. As we noted in connection with contraceptive remedies, he contents himself with giving an account of what he has read:

Alzabo, as it is stated in the *Book of the Sixty Animals*, is an animal which, living in the deserts of Arabia, is of great medical use. . . . It is said that hair taken from the neck of this animal, powdered, mixed and burnt with pitch, forms an anal ointment which cures the sodomite of his vice.[66]

We are here in the realm of magical and analogical thought: an animal traditionally linked to homosexuality can remedy the disadvantages involved by that kind of behaviour. Albert records the recipe without comment. In the same way, he ascribes contraceptive powers to the 'calcaneum' of the weasel, but does not imply that this animal has any particularly abnormal sexuality.[67] Another animal's behaviour attracted attention. The dog appeared quite naturally to be an especially good figure when it came to discussing an accident that could occur during the performance of the sexual act, and which modern medicine calls 'penis captivus'.[68] Medieval doctors set little store by this mishap. Moralistic literature, on the other hand, saw it as a punishment for those wicked people who had illicit carnal relations or who transgressed a prohibition. An offence committed in a sacred place, or even some distance away from it, was also punished. This scenario was very highly elaborated in the second half of the thirteenth century. This type of story could also demonstrate the public shame of a couple, as well as the collective punishment of a whole population's unbridled sexual appetite, as we find in Saxo Grammaticus.[69] However strange it might appear, an anecdote on this topic occurs twice in the *Livre du Chevalier de la Tour Landry pour l'enseignement de ses filles*.[70] Here is the most elaborate form of this edifying story in Caxton's translation:

Hit happed in a chirche on an even of oure lady, one that was called Pers Lenard, whiche was sergaint of Candee on the night delt fleshely with a woman on an auter, and God of his gret might wolde shewe that they dide evelle, tyed hem faste togedre dat night and the morw all day in the sight of the pepill that come thedir into the towne; and all the contre there about come downe and sawe hem. And thei might never parte, but were fast like a dogge and a biche togedre, that night and the morw all day until the tyme that the pepill yode a procession about for them to pray to God and that orrible sight might be ended and hidde and atte the last, whanne it was night thei departed. And after the chirche was halowed or ever there were saide therein ani masse. And they that dede the dede were ioyned to penaince to go naked afore the procession thre sondayes beting hem self and recordyng her synne tofore the pepill. And therfor here is an ensaumple that no body shudde do no suche filthe in the chirche, but kepe it clene and worshipe God there inne.

This is a particularly demonstrative example, giving support to a vigorous moral pedagogy. Does it not really concern the psychology of young girls alone? It seems rather that this story demonstrates the perennial nature of a masculine obsession found in numerous civilizations. The prohibitions laid down by the censor give us an insight into his psychology. As for the form taken by the story, it could correspond to a period of reactivation of the theme, and consequently, it should be included among the component elements of the psychology of medieval man, as one of the forms assumed by his fear of women.[71]

This timid excursion into the extremely rich world of the medieval bestiary leads us to consider another of the 'unnatural pleasures' which, following the Thomist classification, comes also from a corruption of the soul: bestiality. Zoological works traditionally grant an important place to the description of the bear, an animal to which they ascribe sexual behaviour close to that of humans. Vincent of Beauvais emphasizes the fact: 'Bears do not make love like the other quadrupeds, but can embrace each other mutually, like human beings.'[72] Indeed, the bear makes an appearance in the mythical representations of love between different species. William of Auvergne reminds us that, over and above its sexual behaviour, the bear has a face which resembles that of humankind.[73] To illustrate this, he cites the Saxon legend which has the wife of a knight indulging in a long period of cohabitation with a bear; from this 'liaison' were born several children similar to human beings but having a bear's face. Omnipresent in folklore, the bear has the particular feature, according to William of Auvergne, of producing a semen compatible with human modes of reproduction. It is not within our purview to tackle the extensive literature, both scientific and mythical, devoted to monsters. We will content ourselves with quoting Vincent of Beauvais:

Aristotle. Sometimes when the movements of the active virtue are weak and do not manage to vanquish matter, their general character remains and an animal is formed; but this may have the head of a boar or a bull, or else in a similar way a calf with the head of a man is engendered, or a lamb with the head of a bull. This type of monstrosity sometimes occurs in this way, that is, by means of coitus between different species, or by means of an unnatural type of copulation.[74]

The beginning of the quotation does indeed refer to the *Generation of Animals*, but the end differs from it; while Aristotle alludes to the

FIGURE 4.4   *Pasiphae and the bull. Ms Fr. 137, fol. 102v. (Bibliothèque Nationale, Paris)*

particular case represented by the generation of the mule, in the passage referred to by Vincent of Beauvais, he refutes the possibility of procreation between different species: 'The production of all these monsters is due to the causes I have named; at the same time, in no case are they what they are alleged to be, but resemblances only.'[75] The principal obstacle lies in the length of gestation which is different in humans, sheep, dogs and oxen, Vincent of Beauvais's conclusion diverges from Aristotelian theory; for him, there may be not only procreation between man and animal, but also the generation of a monster 'by an unnatural kind of copulation'. Medical literature hardly ever tackles this problem: it alludes merely to monstrosities which, due to a surplus of matter, have too many limbs. The Salernitan texts, however, quote the case of Lombard women who sometimes engender, in addition to a normally constituted child, an animal called *arpo* or *arpa*, resembling a toad.[76] The explanation does not involve sexual relations with an animal. The monstrous creature is produced by a surplus of menstrual blood, an impure residue. The Salernitan authors recall that in derision children call their brother *arpo*. This remark from the world of folklore should be set in a completely different context from that of bestiality. Some texts indeed point out that the monstrous little creature may be born after the ingestion of 'herbs'; Giles of Corbeil is even more explicit: wormwood has the virtue of purging menstrual blood, expelling the inanimate foetus and 'the brother of Salerno'.[77] As it is never specified that the rejection of the monster takes place at the same time as normal delivery, one is allowed to surmise that this legendary phenomenon refers to ways of inducing abortion. The 'Salernitan brother' would merely be a foetus judged undesirable, especially if formed at a time too close to the previous pregnancy. The imaginative evocation of monstrosity and the scientific explanation would then serve only to mask a more unacknowledgeable reality.

Although mentioned constantly in the works that make provision for a tariffing of penances, bestiality lies outside medical discourse; so it seems to belong strictly to the realm of sin, of the soul's corruption. Yet this sexual behaviour could have been included in the doctor's domain. Muslim Spain gives us an example of this happening. At the end of the tenth century, ibn-Yûlyûl recounts, in amusing style, an anecdote involving the Andalusian doctor Yahya ben Ishâq and a peasant.[78] The latter arrived at the practitioner's home mounted on an ass and howling with pain. He was suffering, so he said, from a 'tumour in the urethra' and from the retention of urine. After rapid surgery which caused a momentary black-out, the pus came out and the patient began to urinate. The cause of the obstruction was recognized in the unusual presence of a seed, whereupon the doctor called the peasant a 'corrupt man' and forced him to confess to sexual relations with an animal. Ibn-Yûlyûl concludes by praising the doctor's intuition and exceptional competence. At the same time as adding to the renown of Yaḥya ben Isḥâq, the story casts derision on peasant morals; but it may also be read as a sort of *exemplum* designed to show the dangers of certain types of sexual behaviour. By way of anecdotes of this type, medical hygiene links up with ethics. Literature written in Latin provides us with no such example: the horror aroused by bestiality, a sin against nature, wins out over the desire to warn and to cure.

In the medieval mentality, relations with the animal often act as a transition towards the supernatural world. The *De Universo* of William of Auvergne mingles together bestiality and demonic intervention. Indeed, it is in the chapter that he devotes to incubi and succubi that the Bishop of Paris sets his discussion on the behaviour of the bear;[79] when supernatural creatures decide to fertilize a woman, they choose to do it through the intermediary of a seed capable of creating life in a human womb. From this point of view, the bear presents the greatest advantages. William of Auvergne is talking principally from within a demonological framework; in a text closer to science, the Polish scholar of the thirteenth century, Vitelo, also asks himself how incubi and succubi go into action.[80] The first part of his text, which is presented in the form of a letter, deals with images falsely perceived by the mentally ill, such as maniacs, melancholics, epileptics or apoplectics: these sufferers believe they can see demons, but these are imaginary. Vitelo then describes the visions resulting from the particular conditions of visual perception. After this discussion, which presents a sketch of the distinction between hallucination and illusion of great interest for the history of psychiatry, the Polish scholar enquires into the demonic phenomena that are irreducible to an error in perception. He raises the traditional question of the nature of demons, of their corruptibility, of their capacity to reproduce. While their nature is not that of an animal, these beings may take the form of one by artifice. More often than not invisible to the naked eye, they are perceived by the *phantasia*, the faculty which, placed in the foreparts of the brain, seizes and retains images. How then is one to explain that these imagined forms, that is, those imprinted on the brain, can give a noticeable pleasure to the sense of touch and lead to the sexual act? In the reason put forward by Vitelo, we again come across the importance granted by medieval scholars to the imagination: it 'operates to a great extent during the act, and the spirits, moved by the imagined form, suddenly run throughout the body to aid coitus'.[81] As the motor of physical processes and itself an intermediary between mind and matter, imagination allows demons to couple with human beings. We have seen the same explanatory model enabling the oddities of heredity to be accounted for; in the case of adultery, for instance, the image of the husband remembered during the act makes the child resemble the absent man and not its natural father.[82] Vitelo adopts essentially the point of view of perception; he devotes no particular discussion to the reproductive process. Nonetheless he recalls the origin of the enchanter Merlin, born from an incubus father. He also quotes a case that occurred in Padua in 1265, but without mentioning any offspring: a woman confessed to a priest that she had lain with a 'horned goat'; the demonic nature of this animal seems to be attested by the fact that the animal disappeared immediately after the act and could not be found.[83] For all these phenomena, the Polish scholar tries to give a rational explanation.

It may seem astonishing that, although he refers in the first part of his letter to various mental illnesses recognized in the Middle Ages, Vitelo does not mention the incubus that lay directly within his field of investigation. Should one therefore deduce that he does not consider the phenomenon to be a real nosological entity, and that on this point he diverges from medical tradition? According to him, there may be either false perceptions resulting from serious mental disturbances, or else the influence on the imagination of supernatural beings. But the incubus is not considered by the Polish scholar to be a particular illness. Ever since Caelius Aurelianus, the picture of this disturbance was nonetheless part and parcel of the nosography written in Latin.[84] It was thus considered to be a fantasy which occurs during sleep, either as the consequence of a chill in the brain, or of a difficult digestion, or else as an effect of a harmful vapour rising up to the brain. The sleeper experiences a sensation of oppression and suffocation which may lead him to believe that another body is weighing him down and suffocating him; this leads him to shout and throw himself out of bed. The aetiology of this phenomenon – which modern medicine identifies with a respiratory anomaly – brings into play no sexually linked cause; in the same way, the symptoms are deprived of any erotic sensation: one's only impression is of suffocation. The doctor gives a reassuring interpretation of incubi:

Incubus is a fantasy which during sleep oppresses the body, causing movement and speech. *Incubus* is the name of a demon: some people think that when the incubus moves towards the human body, especially if the man is lying on his back, by reason of a corrupt influence it weighs the body down and leads to belief in suffocation. When this happens to babies in arms, they frequently choke to death, because they cannot stand such a great corruption; that is the opinion of the theologians. But the vulgar think that it is an old woman treading down and compressing the body: far from it. Doctors have a better opinion.[85]

Constantly, medieval medicine opted for a scientific explanation of supernatural phenomena. It did not hesitate to describe in great detail theological positions or popular beliefs that it could then disprove them rationally. Every demonological explanation of incubi and succubi was thus rejected. But the phenomenon was said to be caused by a sensory illusion whose origin was located in the brain, and not by a mere disturbance in breathing. The incubus was still a fantasy, but not an erotic one.

The problem of incubi and succubi lay within medical competence, since it gave one an opportunity to use a rational language and to combat beliefs that might arouse fear. On the other hand, scholars avoided alluding to sins 'against nature'. Their omissions or their condemnation, placed limits on what they considered to be admissible from the point of view of nature: if, in their description of gonorrhoea and nocturnal emissions, they admitted masturbation into their field of investigation, they largely excluded homosexuality. The opportunities provided by Avicenna of tackling relations between people of the same sex were most often neglected. The authority of the Arab scholar was not enough to overcome their reticence. As for sexual practices between people and animals, they lay completely outside medical scrutiny.

5

The Exposed Body

From the vast domain of medieval pathology, we will cite only a few examples capable of shedding light on the representation of sexual behaviour. We will leave to one side gynaecological illnesses which do not seem to threaten the life of the couple, and concentrate on physiological dysfunctions and disorders that prevent or on the other hand require sexual activity. Among the latter, we have already referred to certain sexual irregularities such as satyriasis and priapism, which doctors tended to link to the phenomenon of involuntary emission. We will also enquire into the diseases that were traditionally associated with a life of 'debauchery'. With the help of knowledge derived from modern medicine, we will attempt to find out which venereal diseases may have raged in the Middle Ages, and we will thereby attempt to interpret certain beliefs relative to the transmission of leprosy, beliefs for which a purely religious explanation does not seem to us to be sufficient.

UPSETTING A FRAGILE BALANCE

The medieval physiological system was so coherent that no disorder was left unexplained. Sexual appetite and the reproductive faculty brought into play numerous agents ranging from humoral matter to the imaginative virtue – agents which often behaved in accordance with an immutable order. The absence of one of them or an aberration in its action was enough to break the chain of processes and to entail an imbalance that could affect the body as a whole. Medieval medicine was thus in a position to give an answer to the questions that society asked of it, in relation to impotence for example; it also attempted to treat as patients women who were considered by public opinion, especially at the end of the Middle Ages, to be possessed.

Impotence

The problem of impotence was, principally from the twelfth century onwards, the subject of prolonged thought on the part of theologians and specialists of Canon Law.[1] The medical explanation of physiological mechanisms that we set out at the beginning of this book played its role in the establishing of procedures for the annulment of marriage. Pierre Darmon's book demonstrates the main lines of this influence sufficiently, and there is no need to go over the same ground.[2] The way the congress proceeded, for all the shocking aspects it has for a modern mind, sprang nonetheless from the desire to obtain a technical answer to the problem of impotence. It was too difficult to separate strictly a wife's legitimate claims from her desire to hurt her spouse for the authorities not to surround themselves with the most stringent precautions. Frigidity and impotence have at all times constituted grounds for the annulment of marriage: the contract imposed a debt (*debitum*) which individuals afflicted with infirmity could not pay off. The progressive recognition of the marriage of old and sterile people to some extent established the right to pleasure, with a degree of boldness that is astonishing, for old people, as Huguccio puts it, were thought capable of copulating 'by nature, by artifice or with the help of medication'. On the other hand, the Church could not tolerate the slightest breach of the debt in the case of a couple capable of having children, since the quest for pleasure risked remaining the sole purpose and the sacrament would have been imperilled. The question we wish to raise is the following: did the demands of Christian marriage and the conditions required for it to be dissolved lead to a modification in medical discourse, and a new preoccupation on the part of writers?

Greek and Arabic sources mention the disorders that place obstacles in the way of the performance of the sexual act or of procreation, but, except in cases of congenital malformation or accidental mutilation, these sources do not put forward any codification or classification of irremediable obstacles. The term impotence (*impotentia*) is hardly ever used; the distinctions introduced by theologians between *impotentia respectiva, naturalis, accidentalis, certa, dubia*, etc., and the reliance on the concepts of 'frigidity' or 'arctitude' are a fairly clear demonstration of the problems one faces in bringing together under one word infirmities that have different characteristics and which include a psychological component. There was no treatise on legal medicine in the Middle Ages. The two Arabic works in most demand, the *Pantegni* and the *Canon* of Avicenna, gave no clear definitions on the subject.[3] Constantine the African includes a chapter on 'Those Who Cannot Unite', but the only

topic treated in it is that of evil spells. Doctors could nonetheless find
in these sources, behind the words or phrases designating sterility, or
else defective, weak, or diminished coitus, a description of the principal
physiological deficiencies. Anomalies in the organs such as an 'obstruction
in the womb' or narrowness in the female were described at great length,
at the same time as norms for the male member were proposed. As we
have already seen, another topic was anatomical incompatibility between
partners and discordance in their rhythms. Medieval authors took over
this information without changing it substantially. To every problem,
medicine attempted to bring a solution: its vocation was to explain and
to cure, not to remove all hope or to condemn. Nonetheless, a new
preoccupation emerged from the fourteenth century onwards. The treatise
on male and female sterility attributed sometimes to Arnald of Villanova,
and sometimes to Raymond of Molières,[4] is an example. Concurrently,
general works, types of manuals meant for students' use but also for that
of practitioners, paid more attention to signs that could show that the
generative function was not working. Bernard of Gordon in his chapter
'De Paucitate Coitus, sive de Frigidis et Maleficiatis, de Sterilitate Viri'[5]
still hardly diverged from the outline proposed by the *Canon* of Avicenna.
The *Rosa Anglica* of John of Gaddesden demonstrated a real desire to
classify signs, divided up according to type of cause.[6] If male sterility
came from an external cause, that is, from an accident, from a wound
in the genitals or a hernia,[7] an affliction frequently mentioned and
treated in the Middle Ages, it could be revealed by questioning the
patient. Internal causes were clearly a greater source of difficulty when
it came to making a diagnosis; the signs were recognizable from 'notable
infections of the principal organs'. Evidence for 'paralysis of the penis',
one of the causes of an absent erection, was provided by the cold-water
test. One could recognize, by sight and touch, cold in the testicles,
characterized by a relative lack of hair, tight veins and the coldness of
the skin. Other signs of the same type revealed a complexion that was
too hot, often the cause of premature ejaculation, or temperaments that
were over-abundant in dryness or moistness, which made the man
incapable of producing sperm of the right quality. The characteristics of
such sperm, its consistency and abundance – or lack of them – were
clearly valuable signs of the nature of the obstacles to reproduction. In
1418, the *Philonium* of Valescus of Tarentum presented an even more
detailed classification of the signs of a sexual problem (*defectus coitus*).[8]
His chapter on male problems begins by recapitulating the three elements
necessary: spirit, matter and sexual appetite. In the absence of these
three elements, one should surmise 'frigidity' or an 'evil spell'. If only
one of these agents is missing, the cause may be isolated and the problem

can be treated as one of a 'diminution' or a 'failure', the different degrees of which should be specified. Among the agents responsible, one notes the imagining or remembering of 'something irrelevant'. As in John of Gaddesden, the signs are divided up in accordance with their different types of causes. Valescus of Tarantum complements his theoretical knowledge with evidence from his practical experience. With regard to the questioning which allows the presence of an external cause to be recognized, he records the case of one of his patients who, married to a 'pretty woman', became impotent after suffering a war-wound. Likewise, in his analysis of the anatomical anomalies recognizable to touch and sight, he reports having examined men whose penis was the size of that of a two-year-old boy. As far as internal causes are concerned, Valescus of Tarantum cites the cold-water test and the same symptoms as John of Gaddesden. The signs of female sterility are the object of a similar classification, depending on the anatomical deficiencies (obstruction or narrowness) and the peculiarities of the complexion.

These attempts at classification constitute an embryonic science of legal medicine, and they bear witness to the increasingly common intervention of doctors in procedures for the annulment of marriage. The evolution of Canon Law is not sufficient to account for these new obligations; one must also remember the development of the medical profession within medieval society. Recourse to the expertise of a practitioner became general in all lawsuits and offences that could require his art. The *Register of Civil Causes of the Episcopal Officiality of Paris* includes two examples of a marriage being annulled for the year 1385.[9] In each of these cases, the two masters in medicine and the surgeon, appointed to act as experts in the case, concluded that the husband was impotent. The account of the procedure gives us hardly any details on the way the examination was carried out:

Today we have declared null and void the marriage between Jean Carré and Jeanne la Houdourone de Lagny, thirteen months after the marriage contract, because of the frigidity, incapacity (*inhabilitas*) and impotence of the husband. We have heard the report of Masters Guibert de Serseto and Guillaume Boucher. Masters of Medicine, and of Michel de Pisis, our sworn surgeon, who reported to us that they examined the aforementioned man and that they found and deemed him incapable of knowing a woman. And we have heard the sworn oath of the said woman.[10]

It was in the fourteenth century that the expertise of lettered practitioners seems truly to have become part of everyday manners and customs, acting as a complement to or a check on the examinations carried out by midwives. Guy of Chauliac (died 1368) emphasizes the delicacy and importance of such a task, at the same time as he alludes to the notorious procedure that was to be called the 'congress':

Truth to tell, since justice has adopted the habit of asking the doctor for an examination, the form the examination takes must be described. Once he has obtained permission from justice, the [doctor] must first of all examine the complexion and structure of the reproductive organs; then he must go to a matron used to such [procedures] and he must tell [the husband and wife] to lie together on several successive days in the presence of the said matron. She must administer spices and aromatics to them, she must warm them and anoint them with warm oils, she must massage them near the fire, she must order them to talk to each other and to embrace. Then she must report what she has seen to the doctor. When the doctor has been informed, he must bear testimony in all truth before justice. He must however beware of being deceived, because numerous frauds are habitually committed in such cases, and there is a great peril in separating those whom God has joined together, if it is not for a very just cause.[11]

By this often-quoted text, the efforts demonstrated by writers to help in discovering rapidly the visual signs of some disablement are justified. The doctor's intervention consists essentially in examining the genitals; as for the performance of the act, the practitioner relies on the matron's report, which he must interpret in front of the judges.

It is impossible to deal with the theme of impotence without alluding to evil spells. On this question, authors reacted differently depending on their personality, some of them showing a certain scepticism about the reality of supernatural intervention.[12] One of the signs of impotence caused by an evil spell was that the man could have sexual relations with women other than his wife. In the fifteenth century, Nicholas Falcucci said that supernatural intervention was proven in the absence of any other discernable cause. Medical discourse stressed the role of imagination in explaining these wicked actions, as we have emphasized on several occasions. By including in his commentary on Avicenna's chapter on sodomites an anecdote on the knotting of an aglet which prevented a knight from knowing his wife carnally, Jacques Despars suggested another type of explanation:[13] impotence was frequently nothing more than the result of latent homosexuality. Confirmation seems to be given in the remark that, in this case, it was a man who tied the knot and cast the spell so that the knight 'could not carry out coitus perfectly with his wife'.

The principal source on the problem of evil spells is still the *Pantegni* of Constantine the African in his chapter 'On Those Who Cannot Unite'.[14] From the list of evil instruments and substances, we will pick out beans:

There are other evil spells brought about by the seeds of beans, neither softened in hot water nor cooked over a fire; the worse spell consists in placing four seeds in the bed[15] or on the path or at the entrance or below.... If the spell is caused by beans, it must be cured by divine aid rather than by men.

More frequently than not judged favourable to erection, like other flatulent substances, the bean, whose shape traditionally reminds people of the testicles, is here deemed to cause impotence, in accordance with an ambivalence current in magical thought. It must be pointed out that the evil spell works if the seed stays raw: this detail has its importance. Within a long tradition relating to this forbidden food, one may compare the belief recorded by Constantine the African with what was said, apparently for the first time, by Clement of Alexandria.[16] According to this second-century author, the reason for the ban imposed by legislators on the consumption of beans was allegedly that it made women sterile. The excessive value attributed to this raw foodstuff in medical history and its association with evil spells can easily be justified by the probable presence of favism in the Mediterranean regions that Constantine the African travelled through.

The hysterical malady or the suffocation of the womb

And yet, except for the fact that it is a 'functional' disorder, without concomitant organic pathological changes, it defies definition and any attempt to portray it concretely. Like a globule of mercury, it escapes the grasp. Whenever it appears it takes on the colors of the ambient culture and mores; and thus throughout the ages it presents itself as a shifting, changing, mist-enshrouded phenomenon that must, nevertheless, be dealt with as though it were definite and tangible.[17]

These remarks by Ilza Veith describe perfectly the confusion felt by the historian who is faced with the task of tackling the theme of hysteria. Modern psychiatry, which includes within this word 'three groups of facts',[18] does not make his or her task any easier. When historical hysteria is referred to, it is generally the conversion form that is being thought of, but the object under investigation has the most indistinct contours. Let us specify that we will here be investigating the history of the concept and not the nosological reality that may be hidden beneath other ancient entities.

The medieval phenomenon was called *suffocatio* (or *prefocatio*) *matricis*, the Latin translation of an expression found in Greek and Arabic. It must be stated straight away that, as opposed to what is often suggested, the famous passage in the *Timaeus* in which Plato compares the womb to a living creature – an animal[19] – had only a slight influence on medieval medical ideas. The Hippocratic idea of a uterine displacement, which may have sprung from the observation of prolapses but which also referred to the 'hysterical bolus', was known as early as the high Middle Ages, especially through the *De Mulierum Affectibus*.[20] Because of this displacement, the womb came into sympathy with the upper parts of the body; this explained suffocation and sensory disturbances, and justified the use of fumigations: to push the womb downwards, one made the patient inhale through the nose foetid-smelling substances, while aromatic fumigations were applied to the vagina. On this point, the high Middle Ages also knew, through the *Gynaecia* of Moschion, of the opposite opinion of Soranus, who judged this utilization of fumigations to be useless.[21]

For the medieval doctor, suffocation of the womb was without any possible doubt a disease caused by chastity. The two sources of Greek origin we have just referred to recorded the case of widows and virgins. The *Viaticum* of Constantine the African, by spreading an approximate version of the Galenic interpretation, fixed this idea even more firmly, and definitively, in people's minds:

The cause of this passion is the abundance of sperm or its corruption. It occurs when women are deprived of union with a man: the sperm increases, becomes corrupt and begins to resemble a poison. Widows suffer particularly from it, especially if they have had several children. Likewise, young girls suffer from it when they reach the age [of puberty] without knowing any man, for the sperm accumulates to be expelled, just as in men, as the action of nature requires. When the woman has no commerce with man, the sperm accumulates and there is born from it a smoke which rises to the diaphragm, for the diaphragm and the womb are linked, and, as the diaphragm is linked to the upper parts of the body and the instruments of the voice, suffocation occurs.[22]

The interpretation of *Trotula* is no different; the suffocation of the womb is caused by a retention of seed in virgins and widows: 'In them abounds the seed which nature wished to draw out by means of the male.'[23] Medicine constantly underlined the danger represented by an obstacle placed in the way of a natural function. That the disease was caused by the retention of menstrual blood or of seed could only reinforce a belief in the harmfulness of these substances. While the sexually active woman could contaminate man without herself suffering – as we shall see with regard to leprosy – the chaste woman lost her immunity: the venom turned against her organism and led her to the verge of madness or death. To be sure, medical texts did not make this kind of conclusion explicit, but some of their observations could open the way to such an interpretation on the part of the more misogynous readers.

The *Canon* of Avicenna spread the idea of a complex aetiology of the disease and put forward an attempt at a synthesis of the Hippocratic and Galenic data.[24] The multiplicity of causes accounted for the variation of symptoms from one woman to another. There was an 'organic' (*officialis*) form, due to the womb's rising as an effect of the enclosed matter pushing it upwards. The other cause was called 'material', because it was the matter itself which spread out towards the brain across the membranes, vessels and nerves; in this case, uterine displacement was not necessary. If the retained substance consisted of menstrual blood, there was a great danger that madness might set in, for menstrual blood contained a mixture of the four humours; when melancholy was overabundant in it, its spreading to the brain could lead to serious disorders of the reason. The spermatic form was nonetheless the most pernicious; it gave rise to the severest suffocations and spasms, going as far as fainting fits and apparent death. It was also preceded by a greater sense of oppression. It was after all logical that the substance which could contaminate man during sexual relations should also be the most harmful for the woman if it was not expelled.

The symptoms of the illness often led to its being confused with epilepsy, apoplexy or lethargy. The principal difference lay in aetiology, but certain peculiarities were discernable in the signs. The hysteric did not foam at the mouth to the same extent as the epileptic; as opposed to apoplexy, *suffocatio* did not entail fits of groaning, but rather a state of hebetude; lastly, the illness could be distinguished from lethargy by the absence of fever and a great changeability in the colours of the face, going from pallor to yellow and red. Avicenna distinguished with great clarity between the different phases of the illness. Difficulty in breathing, headache, listlessness and oppression were among the premonitory signs. The succeeding phase could include drowsiness, a flushed face, mental confusion, grinding of the teeth and convulsive contractions. Finally, the state of crisis (especially if a spermatic form was involved) was characterized by a fainting fit, loss of voice and the legs rising up towards the chest, while a film of sweat appeared on the body. We can thus recognize the classic picture of hysteria. The outcome of the crisis emphasized the possibility the illness had of being converted into another; the term of *conversio* was used for this purpose by writers at the end of the Middle Ages, especially by Jacques Despars in his commentary on the *Canon* of Avicenna. Depending on the prevailing humour, there were vomitings, migraines and stiffness, but the symptoms could also be transformed into angina, an 'abscess in the chest or back', disturbances in vision, or even madness.

The spermatic form was considered to be the consequence of a burdensome chastity, and so the expulsion of the retained sperm appeared to be necessary. To obtain this, massaging the genitals with the help of ointments was recommended; Galen described a case of spontaneous cure obtained in this way.[25] Doctors in the Middle Ages recommended the same practices: Arnald of Villanova recommended massage for widows and nuns, and the placing in the vagina of substances designed to lead to assuagement.[26] Here is the advice given by John of Gaddesden:

If the suffocation comes from a retention of sperm, the woman should get together with and draw up a marriage contract with some man. If she does not or cannot do this, because she is a nun and it is forbidden her by her monastic vow or because she is married to an old man incapable of giving her her due, she should travel overseas, take frequent exercise and use medicine which will dry up the sperm. . . . If she has a fainting fit, the midwife should insert a finger covered with oil of lily, laurel or spikenard into her womb and move it vigorously about.[27]

Marriage was the best remedy for chronic suffocation. In cases where this was impossible and other procedures met with failure, masturbation had to be considered, whether practised by the patient herself or by the midwife. For the seed to be expelled, pleasure had to be experienced. According to Avicenna the cure was efficacious only if the sensations of coitus, that is, pleasure and pain, were felt. Medieval doctors generally omitted to refer to this aspect of the treatment when recommending recourse to manipulation; yet it was evident that female 'seminal' discharge accompanied orgasm. Medical discourse repeated this fact again and again on other occasions. Jacques Despars was one of the rare writers to pose the religious problem clearly: 'This practice is only excused and exempted from being counted a sin against nature on condition that it is necessary in order to prevent death. On this point one must consult the theologians so as not to lose one's soul by taking excessive care of the body'.[28]

Gynaecological diseases often required manual relief. In his *Commentary on the Sentences*, Albert the Great stated the distinction between *manus polluens* and *manus medicans*:

The hand which defiles leads to flabbiness or sodomy, but the hand which cures does not, as we say for women who suffer from a fall in the position of the uterus: it is prescribed that this be put back in place by means of the hand and we say that the hand does not defile nor corrupt these women, but rather cures them.[29]

May the same reasoning be applied to the treatment of suffocation of the womb, in which 'defilement' does indeed have to play a part? Doctors in the Middle Ages seemed to suggest as much, but it must be pointed out that they mentioned either individual manual relief, or else the hand of a midwife. At no time was the intervention of a practitioner of the male sex envisaged. This leads us to ask about the conditions of the gynaecological examination when it was practised by a man, which was, furthermore, something exceptional in the Middle Ages. Paul Diepgen does not date manual intervention earlier than the sixteenth century.[30] The examination was carried out with a probe, a wad, or even a mirror. One may recall that, concurrently, the external female genital organs made hardly any appearance in anatomical or pathological descriptions: more often than not, the only organ dealt with was a hidden one – the womb.

If evil spells were sometimes mentioned in order to account for cases of impotence caused by a disordered imagination, the hysterical malady was entirely 'rationalized' by doctors: female nature was sufficient to explain it, because of woman's unbridled sexual appetite and the imperfection of the substances she produced. It is thus hardly any cause for astonishment that writers did not retain the possibility of a male 'hysteria', although this was suggested by Galen.

LEPROSY AND VENEREAL DISEASES

The knowledge, however rudimentary, that doctors in the Middle Ages had of venereal diseases allows us to reach a better understanding of certain aspects of the sexual life and psychology of medieval people. Establishing precepts, codifying and conditioning their moral life on the basis of illnesses – and especially of infections of the 'pudenda' – was an enterprise whose manifestations were too illustrious throughout the centuries for there to be any need for us to insist on them. The efforts of theologians and moralists benefited from arguments that were all the more efficient as they were supported by medical thinking that was widely usable. Medicine offered a nosological conceptualization in conformity with the demands of the sacred and its textual tradition, especially in the case of leprosy. Nonetheless, one question has to be asked, before passing on to any other consideration: was syphilis present in Europe before the discovery of America?

Which diseases?

The origin of syphilis is a subject of controversy. For some, an endemic form of the disease existed; for others, the sailors of Christopher Columbus were entirely responsible for the raging epidemic that ravaged Europe. The most recent discussion of the subject is that of Mirko D. Grmek, who applies himself to resolving the key question, namely that of the diversity or unity of the treponematoses found across the globe.[31] One can distinguish between four forms or types of syphilitic infections: venereal syphilis, the aetiological agent of which is *Treponema pallidum*; endemic syphilis, transmitted by non-venereal channels, probably due to the same germ; yaws caused by *Treponema pertenue* and pinta, caused by *Treponema carateum*. Yaws and pinta are localized in Africa and South America respectively.

For our purposes, it is necessary to mention two hypotheses admitting a form of syphilis in Europe before Christopher Columbus. According to Hackett, who supports the idea of four distinct nosological entities, venereal syphilis was spread across Europe as early as the Roman conquests.[32] According to Cockburn, who supports the idea of the biological unity of the treponematoses, it was endemic syphilis that raged across Europe, before the great venereal pandemic of the Renaissance.[33]

The opinion advanced by Mirko D. Grmek is radically different. His conclusions are supported by the results of osteo-archaeology. It should be explained that venereal syphilis, endemic syphilis and yaws cause bony lesions that it is possible to discover on preserved bones. The osteo-archaeological research carried out allows traces of treponematoses to be detected on pre-Columbian skeletons coming from more or less every region in the American continent, and on skeletons found in Australia too, but on the other hand, as Mirko D. Grmek puts it, 'No bone earlier than 1500 and bearing the unmistakeable stigmata of a treponematosis has been found in Europe, Africa or Asia.'[34]

Such a proof is capable of refuting the theories of the supporters of the existence of an African or European treponematosis. The entire responsibility for the epidemic of venereal syphilis and endemic syphilis experienced in Europe from the fifteenth century onwards has thus to be laid at the door of the pale treponema, the 'American mutant'. For the medievalist, this conclusion is of some importance, since, unless one goes against the evidence written into these bones, it must be realized that syphilis did not exist in Europe in the Middle Ages. Other explanations must thus be found for the venereal diseases encountered in doctors' descriptions.

Arabic authors distinguished between different forms of rashes or cutaneous lesions, behind which certain critics have tried, a little hastily, to discover manifestations of syphilis, despite the over-general and imprecise nature of the descriptions.[35] Thus, it was thought this disease was what lay behind the dermatosis that the Latin translation of the *Canon* of Avicenna named *sahaphati* (from the Arabic *sa'fat*):

Sahaphati begins with little pimples – non-proliferating, small, and distributed over a certain number of places; then these pimples ulcerate, are covered over with crusts, and turn red. Sometimes a poison comes out of them: in that case, it is called a wet *sahaphati*. Sometimes scaling of the skin occurs. This infection often appears in winter and disappears rapidly. . . . *Balkiati* is a pernicious sort of *sahaphati*, while *aluathi* are melancholy ulcers which appear on the thighs.[36]

The text of the *Canon* says nothing further, apart from mentioning, among the numerous remedies, the use of 'earth of quicksilver'. It is indeed known that mercury was used, at a much later date, in the treatment of syphilis. It is, no doubt, the association of this product and *sahaphati* which enabled the diagnosis to point towards venereal disease, although the examination of the texts provides us with no determining element.

Avenzoar localized this dermatosis more precisely on the face;[37] Valescus of Tarentum, in the fifteenth century, made it a sign of imminent leprosy and situated the rash 'around the nose':[38] '*Sahaphati* are red, livid or white pustules around the nose, accompanied by a redness in the face.' A certain number of dermatoses – even urticaria – could correspond to these symptoms. In the conditions of hygiene of the Middle Ages, an ordinary infection, secondarily infected, could present an aspect that was identifiable only with difficulty.

Reference was also made to the *alchumbra* described by Avenzoar. The picture given of it was just as enigmatic: 'On the loss of sensitivity in the penis. This disease is treated like the deprivation of sensitivity of another limb. Sometimes red pustules, called *alchumbra*, break out on the head of the penis.[39] The pustules may have been those of herpes. As far as the 'loss of sensitivity in the penis' is concerned, is this a reference to the absence of erection or to a real loss of feeling, that is, to an illness that affects the nerve centres as a whole? From such an imprecise description, it is impossible to decide between one of the signs of a more serious illness and the effects of genital herpes.

The way illnesses were described in the Middle Ages hardly enables us to recognize with any precision, on the basis of a single symptom, a definite illness. Numerous abscesses or ulcers of the genitals make an appearance, but in the absence of the other symptoms, one can only suggest a number of different possibilities. Certain of these manifestations were linked to sexual activity: we have indicated that the practice of *coitus interruptus* was associated with the risk of penile ulcers; likewise, writers were prodigal with their remarks on the dangers represented for the man by relations with a woman whose vagina was 'diseased, full of impure and virulent matter, flatulence or other corruptions'. In a general way, descriptions did not trace the exact development of ulcerations nor the clinical picture accompanying them: the final phase was more often than not the only one mentioned.

The most commonplace and widespread of venereal infections was blennorrhagia or gonorrhoea. As we have indicated with regard to self-abuse, the latter term indicates the difficulty of making a diagnosis: male discharge of semen outside the performance of the sexual act must be distinguished from the discharge due to a gonococcal infection. The origin of the germ of the disease is unknown; it seems to have appeared in the course of antiquity, as Mirko D. Grmek suggests:

We prefer to push the birth of the gonococcus back to a date prior to the classical era of Mediterranean civilisation, for it seems to use that such a supposition better accounts for myths about the venereal scourge, for the relatively frequent mentioning of cases of urethritis, for the discharge of semen, for the vulvovaginites in ancient medical treatises, and finally for the biological properties of the germ in its present state.[40]

All the infections we have just referred to in regard to the treatises of antiquity can be found in the medical texts of the Middle Ages, but this still leaves difficulties of interpretation. Complications occurring as a consequence of the illness are beyond the reach of medieval medicine: cases of orchitis for men and of vulvovaginitis and, in particular, polymorphous arthritis for women. We should add that among women, gonococcal infections are more often than not painless and do not attract attention until they have infected the Fallopian tubes. In the Middle Ages, discharges from the genitals caused by infections could easily pass unnoticed. The description of certain forms of female discharge[41] leads one to think that other germs such as *Trichomonas vaginalis* or the fungus *Candida albicans* existed and were active. In the man, smarting sensations during urination are characteristic of gonococcal infections. This difference in symptoms, apart from making the disease incomprehensible, may have succeeded in accrediting the feeling of a profound difference between the sexes: woman could contaminate, but did not seem to be affected herself.

Nicolas-Favre disease will keep us for longer, since it enables a sizeable number of cases referred to by the medical tradition to be explained. This illness, more often called *lymphogranuloma venereum*, is sexually transmissible. The incubation period is short, which makes it easier for the observer to perceive the link from cause to effect, and enables him or her to establish a direct link between sexual relations and the first appearance of the illness. The primary lesions present themselves in the form of one or more ulcers, papules, or vesicles. The first manifestation is sometimes transitory; on the other hand, one may have a more extensive or more persistent ulceration. Several weeks afterwards there occurs the 'inguinal syndrome', that is, an intense inflammation of the ganglions of the groin. Here are formed small abscesses and the skin becomes bluish-red, with multiple sinuses through which there emerges a semi-caseous substance, which however is sometimes frankly purulent or flecked with blood. When the inguinal syndrome develops, the patient is subject to a high temperature. One of the later developments of the illness is *genital elephantiasis* or *esthiomene*, described in these terms by modern medicine:

Vegetations and polypoid growths may develop on the skin surface; recto-vaginal or urinary fistulae may form and the area may break down to destructive ulceration. The oedema may extend from the clitoris to the anus. Elephantiasis in the male involves the penis and scrotum; warty growths, urinary fistulae and non-gonococcal urethritis have been described in these cases. Occasionally elephantiasis of one or both legs may occur.[42]

The development of the anorectal syndrome is manifested either by a fistula, by perirectal abscesses, or by tumour-like swellings. The primary lesion is frequently not recognized and the patient realizes his or her condition at the stage of adenitis. A spontaneous recovery is possible, but at the cost of inguinal scars and permanent granulomatous masses. The primary lesion may become chronic in 5 per cent of cases: ulcers and fistulae on the penis, on the urethra and on the scrotum. It is only later that genital elephantiasis, that is, induration of the tissue and an oedema on the penis or vulva, appears.

Apart from accounting for certain descriptions, lymphogranuloma venereum presents symptoms which could explain the relation established by the Middle Ages between leprosy and venereal disease, a subject to which we will be returning. It is caused by the germ *Chlamydia trachomatis*, that is, it belongs to the genus *Chlamydia* whose different species are at the origin of psittacosis, Hodgkin's disease and blindness in newborn babies. The frequency of ocular infections in the Middle Ages clearly indicates the presence of the trachoma. In short, it would be possible to say that a group of diseases, which included those we have just referred to, plus urethritis and inclusion conjunctivitis, was raging. The same microbe *Chlamydia trachomatis* is responsible both for the trachoma and for lymphogranuloma venereum, something that modern medicine has confirmed experimentally. Passing from the one disease to the other was a frequent occurrence, since the number of blind people proves that trachoma was present in endemic state. The constant association we have encountered between diseases of the eyes and sexuality is then easy to understand. One has to add that, through the absence of hygienic precautions, patients afflicted by blennorrhagia could give themselves ocular infections. One of the most fearful consequences for whole populations was that, during delivery, the infected vagina contaminated the infant who, as soon as it came into the world, could be a victim of ocular disorders capable of giving rise to blindness.

To these venereal infections can be added scabies, a skin disease produced by an animal parasite (*Sarcoptes scabiei*); it leads to lesions on the genital organs (particularly the penis), thighs and abdomen. It is an extremely contagious disease and scratching caused by itching opens the way to infections and doubtless to the inoculation of Hansen's bacillus, responsible for leprosy. In the Middle Ages there was no lack of parasites; the scenes of cleansing of vermin are there as evidence of this. So many vectors were capable – given the promiscuity that was customary in habitat – of spreading a great number of diseases rapidly. Certain manifestations of scabies can in any case simulate leprosy. This ambiguity was present in ancient terminology: the *gaffets* of Bordeaux were the *cagots* but also 'the scabby'; likewise the Castillian *sarna* designated scabies, but also leprosy.[43]

Over and above lymphogranuloma venereum there exist a certain number of infections capable of explaining the various growths on the genitals mentioned by medieval doctors. Their descriptions seem to attest to the presence of *condylomata* or *venereal vegetations* to which Latin authors had given the name of *ficus* and which are represented in medical literature. They are 'soft warts of viral origin which can reach cumbersome dimensions'.[44] They are localized around the anus or genitals. Acuminate condylomata (venereal vegetations, cauliflower excrescences, cockscombs), which are contagious and auto-inoculable papilliform growths, are contrasted with other forms, such as the *chancroid condyloma* of the anus.

As early as a Chartres manuscript of the ninth century, there were listed different types of infections of the anus whose names show the strong influence of Greek: *ragadae, cumdolomatae, agrocordenae*, etc. These growth are generally the size of a 'bean, pea or hazel-nut', but they may also become so protuberant that they seem to obstruct the orifice. If left untreated, some of them may not only cause the anus to swell, but spread to neighbouring limbs, and in particular to the penis, which becomes covered with 'filthy and sordid sores'. A few lines later the author of this compilation abandons Greek terminology to refer to the Latin *ficus*, a term widely used in the Middle Ages, even outside medical literature. A miracle story, composed in 1241,[46] recounts that the pantler of King Philippe Auguste, Geoffroy de La Chapelle, suffered in his youth from a *ficus* which a middle-aged Parisian woman cured him of by advising him to pray to Saint Fiacre, the saint traditionally reputed to be the curer of haemorrhoids.

Abscesses on the genital organs which are not caused by diseases described above could equally well be ascribable to *soft sore* or *chancroid*, the agent of which is Ducrey's bacillus. The period of incubation is between two and five days; a small papule develops on the collum glandis penis in the man, or the labia majora in the woman, becomes a pustule and ulcerates.[47] The soft sore may develop for a period of several weeks and lead to spectacular lesions.

One should not neglect, either, the existence of forms of cancer of the penis and vulva, whose existence is admitted by Mirko D. Grmek as early as 'the most remote periods of human history', with clear descriptions being made in the Imperial Roman era.[48] The case described by William of Saliceto does seem to be a cancer of the penis:

White pustules and fissures and corruptions that are formed on the penis and next to the prepuce, for having lived carnally with a woman who is dirty or poisonous (in other words, windy), retained and trapped between the prepuce and the skin of the penis, and when it cannot be expelled or exhaled it grows and multiplies in the place and when at the beginning one ignores it, it multiplies again and corrupts the skin and blackens it and corrodes it with the substance of the penis which is never afterwards cured. With such corruption there come fevers and a flux of blood and very frequently death.[49]

Leprosy and contamination

The Middle Ages did not draw any clear distinction between venereal diseases and the scourge of scourges – leprosy. Before tackling this question, we should specify the exact form of infection represented by medieval leprosy. There is a problem of terminology. In Hippocratic writings, the term *lepra* is applied to various infections of the skin, whose seriousness is incommensurate with that of medieval leprosy. Lepromatose leprosy, the form that we imagine, appeared under the name of elephantiasis in the Graeco–Latin world. The first description, transmitted by Oribasius, is that of Rufus of Ephesus, who locates the first appearance of the disease in Alexandria, in the third century BC. It seems that leprosy did not become a scourge until the end of the Roman Empire, that is 'after the collapse of ancient civilisation and the brutal change in conditions of life which characterizes the beginning of the high Middle Ages'.[50] Studies of medieval leprosy are not lacking, and documentation on this subject is abundant, which is why we will limit ourselves to recalling the essential features. One notes that the first leper-houses in Europe date perhaps from the fifth century, since Gregory of Tours mentions the house of lepers in France. As early as the sixth century,

FIGURE 5.1   *The falsifiers in Hell. Contained in a manuscript of* The Dívine Comedy
              (*Inf.* **XXX**). *Ms Ital. 74, fol. 89. (Bibliothèque Nationale, Paris)*

the disease was endemic, with all the consequences that entailed. For
the Middle Ages, leprosy was lepromatose leprosy and the terms
elephantiasis and *lepra* became interchangeable as early as the fourth
century. The change was important because it allowed a frightfully
effective reading of the Bible, with a dreadful illustration. God's
punishment was not comprised of just any abstract torment, but the
mutilating disease that was before one's eyes every day.[51]

Lepromatose leprosy was merely the most terrible and most spectacular
development of a disease which included a phase of ambiguity, that in
which spots appeared on the skin. It was at that moment that the person
was generally examined by a jury, composed to begin with essentially
of the leprous, in accordance with a well-known procedure.[52] Doctors
and surgeons, as in the case of other expert opinion, were only called

in from the fourteenth century onwards, that is, at a time when leprosy seems to have distinctly retreated. If the examination proved positive, the patient was excluded from the community of the healthy.

The many guises in which leprosy could appear, and the impossibility of defining the nosological reality with any precision, made this disease a prime subject for theological discourse. Paradoxically, the most exact observation of reality and a lesser resistance to theological and moral influence were found together in the attempt that was made to interpret modes of transmission and causes.

The whole Middle Ages believed that leprosy was extremely contagious, and built lazar-houses, with a system of exclusion that was more or less respected depending on circumstances. Theoretical writings on the separation of the two communities were in any case much stricter than was daily practice. According to doctors,[53] contagion was deemed to be effected by sexual relations, by contact and by breath. Lepers were not to speak to healthy people except to leeward and could touch objects only with a stick. Belief in the contagion of the disease was explained by the authority of the biblical text. Leprosy was a divine sign, a punishment that could strike only the reprobate, and those who did not flee the person on whom God had let fall his wrath would meet with the same punishment. In fact, leprosy is contagious only after repeated contact with a diseased person. In the leprous couple, for example, it was possible for the woman to become progressively immunized and thus show that she was capable of resisting the disease. Furthermore, the length of incubation always exceeds two years. What is most interesting for the historian of attitudes and ideas is thus the system of causality set up to explain something that was beyond the grasp of medieval medicine.

The explanation of the transmission of leprosy by sexual relations has a very ancient tradition. It figures in Indian literature: Suçruta mentions this type of contagion, at the same time as contamination by touch, breathing, or the use of objects utilized by the patient.[54] We thus see an astonishing encounter between this medicine, ancient and distant, and beliefs relative to leprosy in the thirteenth century of our own Middle Ages! Another belief was constantly linked to this disease, that of the inextinguishable sexual ardour of the leper. The fragment of Rufus of Ephesus, transmitted by Oribasius's compilation, describes leprosy perfectly and includes a sentence of great importance:

The doctors who lived a little time before us also established different kinds of this disease; they called it, in its earliest stage, leontiasis, because patients begin to smell bad, their cheeks sag, and their lips thicken; but when the eyebrows swell, when the cheekbones go red, and patients are seized by ardour for coitus, these doctors give the name satyriasis to the disease, which, however, is something other than the infection of the genitals (called by the same name).[55]

What was the influence of this tradition on the way the leper was represented? The certainty of an irremediable condemnation could produce the sexual activity of despair, analogous to that attributed, as late as thirty years ago, to the inmates of sanatoria. The final reason is sociological in nature: healthy people observed without the least favour a minority to which they ascribed vices dreamt up by their own powers of imagination, even their own sexual deprivation. Finally, we may say that theologians doubtless accommodated themselves very easily to a belief that justified the maintenance of a punishment without remission for the lecher.

Numerous authors in the Middle Ages included among the causes of leprosy not only coitus carried out with a leprous woman, but also intercourse during menstruation. This opinion of doctors also figured in writings meant for the use of confessors:

He will ask whether it is a woman who has just given birth, for this is forbidden by the Law, or *in the middle of her period*, which is also forbidden simply because there is in this case a physical danger for the father *because of the risk of leprosy*, and *for her offspring*, for from the seed corrupted foetus is born.[56]

A popular text like the *De Secretis Mulierum* transmitted the same belief and probably put it into wide circulation. It is justifiably easy to explain this by reference to the biblical ban, commonly accepted and interiorized in popular thinking. Nonetheless, certain texts are disturbing. Thus Bernard of Gordon, a doctor from Montpellier, admits a variety of causes of contamination.[57] Leprosy is caused by coitus; a child conceived during a period will be born leprous; the result will be the same if a leper has relations with a pregnant woman; it occasionally happens that the child is born into air which is infected; the abuse of food encouraging the rise of melancholy may have the same consequences. If we take up one by one the causes referred to by Bernard of Gordon, first of all the idea that any one of the diseases that we have been referring to can cross the placental barrier must be rejected. The only infections involved can be those transmitted from mother to child at the time of delivery; Bernard of Gordon would thus be drawing on a medical observation, clearly making a grave error of interpretation. The infections possible are those deriving from the agent responsible for both trachoma and lymphogranuloma venereum. One may also mention genital herpes which today's situation has just brought into the forefront of venereal infections: if the mother is victim of genital herpes, the child, contaminated at delivery, may develop a neonatal primo-infection whose seriousness

varies from simple skin lesion to septicaemia with fatal meningoencephal-
itis.

John of Gaddesden, like William of Saliceto and Lanfranc, admits the
danger of coitus during a woman's period; the threat is more or less the
same when one suspects the woman of being affected by a disease. In
this case, the well-informed author of the *Rosa Anglica* indicates the
precautions necessary: 'If you wish to preserve your organ from all harm,
should you suspect your partner of being corrupted, purify yourself, as
soon as you have withdrawn, with cold water in which you have mixed
vinegar, or with urine.'[58]

The texts are too evocative for one to be content with merely explaining
them by reference to prohibition. We will advance two hypotheses to
find a scientific justification for these remarks. It is evident that the
woman in question was afflicted by a venereal disease. It is useless to
say – and Henry of Mondeville's ignorance proves it – that gynaecological
examinations were not regularly carried out in the Middle Ages. As far
as intercourse during menstrual periods is concerned, one could refer to
the case of an infectious discharge – either permanent or caused by
sexual relations – and interpreted by the partner as a menstrual discharge.
In fact, the doctor had at his disposal only the information given by the
man. Another hypothesis appears more convincing: the woman's periods
are the end of a process that necessitates an important activity on the
part of the tissues, ending with the emission of 'menstrual blood'. It is
probable that on this occasion the germs will take advantage of conditions
particularly favourable to their activity and their transmission, whatever
may be the stage of development of the venereal disease the woman is
suffering from.

That lymphogranuloma venereum, because of the adenitis it causes,
and because of elephantiasis, could easily be interpreted as a form of
leprosy is almost certain. In this regard, an example related by Bernard
of Gordon claims our attention.[59] The person in question was a medical
student at Montpellier; the lover of a countess afflicted by leprosy, he
became 'completely leprous'. It would be possible to imagine that the
unfortunate man, previously contaminated, discovered on his body the
signs of leprosy during or shortly after the liaison. The circumstances of
the adventure struck the imagination of Bernard of Gordon, who,
deceived by a (possible but exceptional) coincidence in time, drew a
false conclusion from this case. We cannot accept this way of seeing
things, since for medieval doctors the venereal transmission of leprosy
was a constantly reaffirmed law and the coincidences – which could not
help but happen in the case of an endemic disease – were not sufficient
in number to explain an immediate causality, the brevity of the period
of 'incubation' that all doctors seem to have admitted.

When medieval medicine speaks of leprosy, what form does the question take? We must beware of thinking that it is always *lepromatose leprosy* that is under discussion. The indefinite form of the disease already aroused the attention of patients and doctors. It could be that the jaundice discovered in newborn babies was considered as a primary form of leprosy, or what we have called *tuberculoid leprosy*, 'the principle manifestations of which on the skin are well-defined and anaesthetic clear marks'.[60] On these children, numerous skin diseases may have led to leprosy being diagnosed.

Our discussion, which takes its cue from a few remarks gleaned from doctors, clearly shows the vulnerability of the child. It has been shown that, for people of the Middle Ages, the child was not protected from the diseases of adults while it was in its mother's womb. It was contaminated at the time of delivery and made a victim of the trachoma or lymphogranuloma venereum; the signs of the disease and therefore the stigmata of God's anger were written on its flesh and its skin. If they survived, what kind of life did those have who came into the world in such conditions? There is a tendency to forget that the Middle Ages was a dreadfully harsh period, and that to the sufferings caused by the disease and the inferiority that it inflicted was added the psychological imbalance entailed by the disapproval of the community.

The alleged immunity of women

The examination of the possibilities of analysing differently what the Middle Ages subsumed indiscriminately under the name of leprosy forces us to judge the texts more carefully. A semiological system had been established, that allowed the complexity of reality to be more or less adequately grasped. The linking of causes to effects is far from satisfying to our modern spirit. For a while, one must accept the gaps and omissions in this synchronic cross-section so as to perceive the intellectual attempts made by the intelligent and always courageous doctors, frequently aware of the derisory weakness of the means at their disposal, and coming up against the radical impossibility of acting effectively. As we have seen, many wrongly interpreted facts led them to postulate that there existed a link between leprosy and venereal contagion and that woman possessed a relative immunity.

Leprosy could be seen by medieval men and women in all its forms, and, in the south-west of France and in Brittany,[61] the *cagots* provided the spectacle of communities within which the infection seemed to be hereditary. In fact, this was an intermediate type of leprosy – presenting at once the characteristics of lepromatose leprosy and tubercular leprosy

– which parents passed on to their children. In a closed environment, intermarriage contributed to increasing the wretchedness of this population. This seemed to provide irrefutable proof that God's curse could strike not only an individual, but an entire community. The nosological way of conceiving things was different from ours and the leprosy of medieval times included, apart from the disease caused by Hansen's bacillus, lymphogranuloma venereum and probably other skin diseases too. In these conditions the idea of a venereal transmission was not a pure fantasy, Adelard of Bath and especially William of Conches – that is to say, the authority who was the most widely read – proclaimed that contagion occurred via the genitals. The idea, clearly expressed in the Middle Ages, had no doubt been developing for a very long time. William's text is particularly explicit: 'Why, if a leper knows a woman, is the woman left unharmed, whereas the first man to know her thereafter will become a leper?' The master replies to the disciple's question by explaining the mechanism of contamination:

The hottest woman is colder than the coldest man: such a complexion is hard and extremely resistant to male corruption; nonetheless the putrid matter, coming from coitus with lepers, remains in the womb. And when a man penetrates her the penis, made of sinews, enters the vagina (*vulva*) and, by virtue of its attractive force, attracts this matter to itself (and to the organs to which it is attached) and transmits it to them.[62]

It cannot be denied that what we have here is an allusion to venereal contamination. The extension of the infection to the parts of the body located next to the penis excludes the possibility of a blennorrhagia, but would correspond closely to lymphogranuloma venereum at the stage of the development of adenitis. Another observation must be made: woman is protected from contamination by her complexion, which does not allow the 'corruption' to circulate. This belief in the immunity of the woman corresponds partly to the materiality of the facts. To resolve this enigma, one must review the diseases, defined by modern medicine, that comprised the medieval entity. As far as true leprosy is concerned, it is possible that men were affected in greater numbers than women. Certain authors have appealed to reasons of a physiological nature.[63] Nonetheless, considerations relating to the way of life of the two sexes provide other equally convincing explanations. Women hardly ever left the home, whereas because of their various activities men were constantly on the move and more prone to come into contact with people capable of contaminating them. That was the situation with regard to leprosy. As far as venereal diseases are concerned, they managed to win powerful support for the idea that women were immune. One must, of course, remember that we are dealing with infections whose incubation period is short, so that the causal link between sexual intercourse and the first symptom of the disease may be perceived. It will be noticed, furthermore, how difficult it is to draw conclusions on the basis of a text like that of William of Conches – and numerous medieval doctors are not far behind him when it comes to imprecision. The process of contamination was alluded to theoretically and the first signs neglected. In the case of lymphogranuloma venereum, the absence of any mention of the first signs corresponds to medical reality. Indeed, it can be seen nowadays that the little painless vesicle or the first ulcer are noticed by only a third of men, and they are more often than not unnoticed by women. Patients start to worry when regional adenitis appears. Female morphology conceals any primary lesion, since this is situated either on the labia, or else on the posterior wall of the vagina. The disease may be cured spontaneously, without leaving any scars. Thus the man, just by having intercourse with a contaminated woman, may contract the disease, while his partner will be left unharmed. It must be confessed that there is plenty here to catch the medieval imagination, preconditioned as it was by a theological discourse that was not particularly favourable to woman. It must be added that the endemic disease caused by *Chlamydia trachomatis* could lead to surprising and unpredictable cases of contamination and apparent immunity. Soft crab in the first stage of development is also concealed in woman, since it affects the clitoris, the labia minora, and the fourchette, and may not be revealed to the man who will see the effects – in his case quite visible – of the disease. Twentieth-century medicine comes up in the same way against great difficulties in the detection of syphilitic crab.

It must be noted, too, that the 'dirty and filthy' women spoken of by the medieval texts must often have been prostitutes in whose interest it was to hide their blemishes. Certain prescriptions bear the trace of the idea that objects could be contaminated by the hands of prostitutes.[64] Apart from the sterility that was ascribed to them in the Middle Ages, which is easily explainable by infections followed by adnexites – at a period when of course antibiotics or any other effective means of defence were unknown – prostitution must have constituted a formidable means of spreading the germs, perhaps indeed because of a certain resistance acquired by these women coming into frequent contact with the agents of infection. So whether one or other of the infections covered by the name of 'leprosy' in the Middle Ages is being considered, we are obliged to recognize that the detection of the disease lay beyond the means of men, or even doctors. Thus popular feelings and medieval science accepted that the woman, in league with lust, could be exempt from diseases the man saw immediately branding his own flesh.[65]

Woman was also capable of resisting another poison that she herself secreted. The harmfulness of the menstrual flow has a long tradition behind it in the history of the human race. It has been seen that in the Middle Ages coitus during this period was deemed to cause leprosy. We have expressed the idea that menstruation could correspond to an increased activity of microbes in the contaminated woman. The possibility of a medical and rational explanation was evidently of less force than the authority of biblical affirmation, as mediated by theologians. Moreover, whatever the reason put forward, the work of the imagination in the Middle Ages transformed woman into a 'machine' capable of producing a certain dose of poison every month. This 'function' established an uncrossable barrier between the two sexes and nourished enduringly the fantasies of *gynophobia*. As for the older woman, after menopause, the poison contained in her body had to find an exit; the venomous humour was evacuated from the body by means of the eyes and these women poisoned animals as well as small children.[66] The female organism was used to the poison that it secreted and this form of immunity was closely related to the question of mithridatism or mithridization.

Acquiring a resistance to poison was a constant aim of men and women of state, especially when political circumstances were difficult. In the Western Middle Ages, it was a girl who acquired this fabulous resistance, and this gave rise to the story of the Venomous Virgin, dear to writers of the thirteenth century. A king who was mistrustful of the growing power of Alexander brought up a girl whom he fed on poison. When the girl was completely venomous, the king sent her as a present to the young Alexander:

And when Alexander had received this fine present and saw the ladies [i.e. her attendants] and the beautiful young girl, agreeable and playing the harp perfectly, she pleased him marvellously and he had to make an effort to restrain himself from going to embrace her; and his temptation was extremely strong. But Aristotle, a clerk at his court, and Socrates, his master, detected the presence of the poison in the girl and would not allow Alexander to touch her. And when they said that to Alexander, he could not believe it, but he was afraid of Socrates, his master, and did not dare to contradict him. Then Socrates had two serfs brought into Alexander's presence and made one of them embrace the girl: he dropped dead on the spot; he made the other one do the same, and he died in the same way, so that Alexander recognised that his master had been telling him the truth. But he did not leave it at that: he made her touch other animals, dogs and horses, who all died on the spot. In this way Alexander was this time saved from death by his master Socrates who was of very great help to him thanks to his extremely penetrating intelligence.[67]

It was the stake that finally allowed the world to be rid of this scourge.

The story was known from India to the West, whither it was transmitted by the Arabs. It spread throughout Europe at the end of the thirteenth century in the form of a perfectly well-organized tale, since it gave the best representation of a dangerous woman that the Middle Ages could imagine. The poison that this slut was capable of getting used to was the secretion of her own organism; the death that she delivered to everything that approached her was the purest, most brutal, but also sincerest expression of an interiorized fear of woman. This fear was the fear of contamination in the different forms we have been discussing. The character of the bawd has enabled us to show the lack of understanding and the hatred that separated man and woman, for he was paralysed by his ignorance and humiliated by the knowledge that he ascribed to the Other. In conclusion, the old woman composing philtres of love or death merely exercised an activity analogous to the transformation that physiology worked in the female organism; this character was simply the last avatar of femininity. We have for a moment left the field of medicine for that of literature, but this was a useful thing to do, for, in the literary tale, the author did not censor his or her fantasies. This exorcism was not always to be sufficient and the man who felt threatened was to put his fantasies into practice. When that time came, witches were sent to the stake.

The most scientific mind of his age, and also the one most favourable to women, Albert the Great, did not always rise above common opinion. He records the case of a little girl from Cologne, aged three years old, who fed herself exclusively on spiders and felt all the better for it.[68] So the little girl could, without any danger, consume the repellent and poisonous spider. What better proof could there be of her complicity, and that of her sex, with poison! One of the 'Miracles of the Virgin' is even more brutal and narrates a sordid story – which seems to have been written so as to be given a psychoanalytical reading.[69] Let the reader judge: a young girl lived with her uncle, surrendered to his advances and abandoned herself to lust. Becoming pregnant three times, she threw her baby down the latrines. Finally overwhelmed by remorse, she decided to commit suicide, shut herself up in a lumber-room that had been abandoned and there tried to poison herself with a spider. Although she had chosen a fine specimen, she obtained no result. She tried three times. In vain. Then she decided to cut open her stomach with a knife. The poison was expelled and with it (though the text does not say this) the poison of lust, since at that instant she repented and the Virgin intervened to save the sinner, who later went on to an edifying end. The intentions are a little too overt, but the author, in his function of preacher, played on the unconscious of the faithful – and struck home, since confirmation can be found, in scientific texts, of this principle of immunity, so fundamental to the representation of women in the Middle Ages.

At the beginning of French literature, the couple Tristan and Iseult acted out the first tragic love story. But this love was already marked by the sign of illness.[70] Tristan sitting on the edge of the swamp, disguised as a leper, declares to King Mark: 'For as long as I was healthy I had a very noble lover, now because of her I have these bumps on my skin [the nodules of leprosy].' When the beautiful Iseult is playing, with her lover, the part meant to dupe the royal suite, she addresses him in these terms: 'Do you think I will catch your disease? Have no fear, I will not.' The woman – the impudent woman – Iseult proclaims her immunity to the whole world. The most beautiful of love stories includes, even if this episode is brief, a creation of the imagination conditioned by common belief. In it, we find the contamination of leprosy or venereal disease as it was referred to by William of Conches. The nosological conceptualization of the medieval world is an indispensible instrument for anyone wishing to discover, without falling prey to anachronism, the key to episodes in literature, and their secret associations. Sexuality and its diseases constituted sufficiently powerful psychological constraints for any attempt at an approach to the medieval world to beware of neglecting the way they were represented.

The fear of women can indeed by explained by other reasons well known to psychoanalysis. Following this method of analysis, one may consider that theological discourse exploited the primitive fears of men. Even if that is written in the Bible, we have left to these influences their role of first cause. What we have been trying to seek out are the secondary causes, namely the way the individual interpreted the accidents that played a part in his or her relations with a particular milieu, in which disease had its own place.

Conclusion

All thinking on sexuality contrasts implicitly – even if it claims not to – two types of discourse. The constraints and restrictions imposed on sexual activity by theology have often been emphasized; it is legitimate to contrast it with the desire to know and explain which inspired medicine. Theologians referred constantly to sexuality, and the penitentials, a catalogue of practices and perversions, may be considered as the earliest systematic exploration of the many manifestations of sexual desire. This quest did not entirely escape being influenced by the culture or indeed the imagination of its authors, and its bookish and often artificial character has been criticized. Confession and the dialogue it involves keep alive throughout the centuries the practice of an exchange of ideas about sexual life and it may well be thought that it aroused the curiosity of penitents by passing certain information on to them. Treatises of casuistry of course took over this role and it has been said of them: 'The *succès de scandale* that they sometimes encountered leads one to think, as Bayle suggests, that this type of work served as much to initiate the public into sexual matters as to edify it.'[1] In the course of this development the writings of theologians were enriched with the scientific knowledge accepted by doctors. The more philosophical than medical direction that thinking about sexuality took with Averroës and especially Aristotle – both preferred to Avicenna – must be underlined. The end of the thirteenth century saw the triumph of the ideas of Aristotle, in spite of the fact that they had been condemned by Etienne Tempier. To quote only one example, his audience was such that the idea of female seed, found in all the doctors of the age, moved into the background, eclipsed by the assertion of the Philosopher who, contrasting matter and form, was at the origin of a great number of scientific commentaries greatly removed from medical reality. The *Nichomachean Ethics* did, it is true, make theological discussions on the nature of pleasure and its recognition as a component part of marriage possible, but a greater number of

Aristotle's writings contributed to implanting among intellectuals and in popular awareness the idea of woman's inferiority. The Stagirite's role may thus be considered pernicious. On the other hand, doctors, following Galen and Avicenna, were the defenders of the doctrine of female sperm and, while seeking for the causes of its emission, they were led to incomparably more subtle psychological analyses. The proof of the quality of the doctors' systematic perspective is provided by casuistry itself, especially by Peter de Palude, followed, much later, by Cajetan, Sanchez and Liguori. Thus, Peter de Palude accepts *amplexus reservatus*, intercourse without emission, for economic reasons, 'unless from this perhaps the woman is provoked to semination.'[2] We will never know – and with good reason – the exact meaning of this expression, but, for the casuist, a closer attention to human reality entails new distinctions and makes a more complete set of conceptual tools necessary – tools which it takes from medicine. Yet the purpose of the two sciences is different: medicine makes efforts to discover a principle that will explain reality; casuistry recovers a particular tool for its own use and, without bothering itself about physiological considerations, dissociates the sexual act into two parts: one is comprised of a permissible satisfaction and the other of an illicit orgasm which sets in motion the process of fertilization. The will to explain creates its object and medicine is there merely to provide an acceptable physiological explanation.

In the course of their history, medicine and theology did not cease to be closely related. Right from the earliest texts onwards, the obstacles to the consummation of marriage preoccupied theologians and many of the medical texts inquired into the cases of impotence caused by a malformation of the genital organs. It may be considered that the procedure of the congress was the outcome of a desire to establish the sacrament on physiological considerations and that it established the doctor as the deciding judge. Throughout its history, theology was to construct its sexual ethics on the quasi-mechanistic conceptions that the medicine of the period put forward. The danger came, indeed, from the excessive nature of a material which could not be entirely controlled, and the word continence should be taken in its etymological sense. Theology could find itself in difficulties when it came to giving the reason for certain physiological phenomena – the superabundance of humour in the woman, for instance. A particularly indulgent theologian such as Albert the Great, for example, presupposed some final cause, while rigorists condemned, and as for those who pursued to the end the attempt to reconcile theology and medicine, they were faced with the problem of discovering the general intention, that is, of finding an expedient to procure satisfaction for the necessities of the body, without the mind and thus pleasure, playing any part in the process.

It must also be admitted that the most well-informed section of the population did not cease to question medicine for an explanation of infirmities and deficiencies, to learn about sexual hygiene and above all to enquire into the causes of sterility in the couple. The end of the thirteenth century, for reasons we have analysed, seems to have inaugurated a bold debate on sexuality, and used all the available texts. The elements of an erotic art were taken from another civilization which practised polygamy and which made the pleasure of the partners into a positive value. Medical texts progressively integrated the elements of an art of love taken from the Arabic world and from Latin antiquity. The liberty of the Arabic texts remained infinitely greater than that of Western texts. Nonetheless, it cannot be denied that at the end of the Middle Ages, people had access to an erotic art and erotic techniques that had nothing to envy other civilizations.

While medicine and theology carefully demarcated their respective domains, they did not for all that, as we have just seen, eliminate every source of conflict. The Church found it difficult to forget that its first preoccupation had to be the soul: 'Insofar as the soul is more precious than the body, the priest must be held in greater honour than the doctor of the body.'[3] Medicine must be only a secondary domain and this science less than any other could escape from the tutelage of the Church. A cure always came about by God's grace and it was less important to follow the advice of the doctor than that of the priest. In every case, the doctor had no right to counsel anything that would place the patient's soul in danger, as Innocent IV reminded people.[4] One of the prescriptions of medieval medicine was the target here: the idea that sexual intercourse outside marriage would bring about the end of the complaint with which the patient was afflicted. At the beginning of the thirteenth century, Vincent of Spain, whose words were to become authoritative, considered it a grave sin for the doctor to indicate, even allusively, the real cause of complaints due to the retention of semen. We have seen elsewhere how sharp the conflicts between the religious imperative and medical knowledge could become when the patient belonged to the monastic community. The theologians proclaimed that sexual abstinence was possible, whereas, as we have emphasized several times, this ideal was difficult to reconcile with medical theories, close as these were to a mechanistic physics which made the movement of the humours irrepressible in certain cases. Without being unaware of the religious implications of its advice, medicine pursued the logic of its system and implicitly left the patient with the responsibility for choice, when saving his body risked endangering the purity of his soul.

In fact, medicine and theology were nourished on the same knowledge, and religious ethics often borrowed from science the arguments that served to control sexuality and its excesses. Both were discourses that explained and controlled behaviour. Both also had the ability to delimit restrictively their field of application. In the name of nature, theology constructed the norm for the behaviour of the ideal couple. Sexuality in other conditions, in other forms, was rejected by a reference to biblical discourse as in the case of Sodom and Gomorrah; it was metaphorically cast out into the darkness of former worlds or into the animal kingdom. In analogous fashion, medicine either ignored behaviour that was not within the the norm or else classified it among pathological cases. The strategy consisted in removing from such kinds of behaviour any possibility of existence in the present, healthy, normal world. The analogy stops there. The Church was to attempt the difficult task of reconciling itself to somewhat over-rigid scientific theories, whereas medicine was to process, with explanatory models that were basically little different, the data of observation supplied by practice and everyday contact with the diversity of human reality, and to try to extend the area of knowledge that would allow control over the body to be achieved.

Notes

Chapter 1   Anatomy, or the Quest for Words

1 O. Temkin, *Galenism: Rise and Decline of a Medical Philosophy* (Cornell University Press, Ithaca, NY, and London, 1973).

2 J. Fontaine, *Isidore de Séville et la culture classique dans l'Espagne wisigothique, Etudes augustiniennes*, 2 vols (Paris, 1959), vol. I, pp. 28–9: 'Thus the grammarians thought that they had the right not only to define and communicate the elementary techniques of expression, but also to extend their authority over the whole field of aesthetics.'

3 Ibid., p. 28.

4 J. Engels, 'La Portée de l'étymologie isidorienne', *Studi Medievali*, third series, 3 (1962), pp. 100–28. Quotation and translation, p. 100.

5

Nephew	= Name, designation
Who is born from the son	= Force (or meaning) of the name
As if born after	= Note (or verbal indication)
For at first the son is born, then the nephew	= Origin or true way of speaking

– ibid., p. 110. Isidore gives the word etymology its own etymological meaning, that of *veri-loquium* ('true way of speaking, in accordance with the reality of things') which he could also find in Cicero and Boethius; see ibid., p. 112.

6 'The most important part of the body is the head, *caput*: it has been given this name because all sensation and nerves take [*capiant*] their beginning thence, and because every principle of life springs therefrom.'

The quotations are taken from the edition by W.M. Lindsay (ed.), *Isidori Hispalensis Episcopi Etymologiarum sive Originum Libri XX* (Oxford University Press, Oxford, 1911); for the present example, see XI, 1, 25. Translation in William D. Sharpe (ed.), *Isidore of Seville: The Medical Writings. An English Translation with an Introduction and Commentary*, Transactions of the American Philosophical Society, n.s., LIV–2 (Philadelphia, 1964), p. 40.

7 Engels, 'La Portée', pp. 112–13.

8 Fontaine, *Isidore de Séville*, p. 44.

9 R. Klinck, *Die lateinische Etymologie des Mittelalters*, Medium Aevum, Philologi-sche Studien (Wilhelm Fink Verlag, Munich, 1970). Here are the two definitions of etymology given by Pierre Elie, as quoted by Roswitha Klinck: (1) 'Est vero "ethimologia" nomen compositum ab "ethimos" quod interpretatur "verum" et "logos" quod interpretatur "sermo" ut dicatur "ethimologia" quasi "veriloquium", quoniam qui ethimologizat veram, id est primam, vocabuli originem assignat' ('In fact, "etymology" is a compound noun made up from "etymos", meaning "true", and "logos", meaning "speech", so that "etymology" means as it were "true speech", since whoever etymologizes attributes to a word its true, in other words its first, origin') (p. 15); (2) 'Ethimologia ergo est expositio alicuius vocabuli per aliud vocabulum sive plura magis nota secundum rei proprietatem et litterarum similitudinem ut "lapis" quasi "ledens pedem", "fenestra" quasi "ferens nos extra". Hic enim rei proprietas attenditur et litterarum similitudo observatur' ('Etymology is thus the explanation of some word by another word or several better-known words in accordance with the thing's special character and the similarity of the letters used, such as "stone" ("lapis") from "wounding the foot" ("ledens pedem"), "window" ("fenestra") from "taking us outside" ("ferens nos extra"). Here indeed attention is drawn to the thing's special character and the similarity in the letters involved is clearly observed) (p. 13, note 17). In the same text is raised the problem of translating a word from another language: 'Differt autem ab interpretatione que est translatio de una loquela in aliam. Ethimologia vero fit sepius in eadem loquela.' ('Etymology is however different from interpretation, which is the translation from one language into another. An etymology is more frequently found within one and the same language').

10 Ibid., p. 18. Thus the author of the gloss *Tria sunt* disputes the over-general meaning given to the word 'etymology' by Isidore: 'Sed largius ethimologia accipitur secundum Ysidorum ut amplectatur etiam expositionem que fit per compositionem vel per derivationem vel per aliam linguam iuxta litterarum similitudinem.' ('But there is a wider sense of etymology according to Isidore: it also includes an explanation that is based on the arrangement of words, or their derivation, or the similarity of their spelling with that of words in a closely-related language').

11 E. Faral, *Les Arts Poétiques du XIIe et du XIIIe siècle* (Honoré Champion, Paris, 1962), p. 65: 'A particular form of *interpretatio* consists in exploiting its etymology (in Latin, *nota* or *notatio*). Classical writers ranged etymology among the *topoi* of invention: their theory is developed by later writers. In *Les Arts poétiques* this point is discussed by Matthew of Vendôme (I, 78) and by John of Garland. A gloss in Latin manuscript 18570 of the Bibliothèque nationale, on verses 41–4 of the *Laborintus*, gives these curious details: "[Interpretatio] aliquando fit per litteras, aliquando per syllabas, aliquando per dictiones. In litteris exemplum, *Deus*: *d*ans *e*ternam *v*itam *s*uis. Exemplum in syllabis, ut *ecce cadaver*: *ecce ca*rnem *datam ver*mibus, Tertia fit per dictiones, ut *materia*: que *mater altera; fortuna* que *forte una*, etc.".'

12 M. Foucault, *The Order of Things: An Archaeology of the Human Sciences* (Tavistock, London, 1970), p. 35.

13 'The breasts, *mamilla*, are so called from their applelike roundness, *mala*, by diminution. The nipple, *papilla*, is the prominence of the breasts, of which those sucking take hold. They are called "nipples" because infants are as though eating pap, *pappare*, when they suck milk. In the same manner the

breast is all distended with abundance, but the nipple itself from which the milk is drawn is small. It is called a teat, *uber*, either because filled, *uberta*, with milk or because moist, *uvida*, filled with the liquid milk as though a grape, *uva*. Milk, *lac*, derives the force of its name from color, since it is a white fluid: in Greek, they say *Leukos* for "white". Its nature is changed out of that of blood.' Lindsay (ed.), *Isidori*, XI, 1, 75–7: translation in Sharpe (ed.), *Isidore of Seville*, p. 43.

14 Engels, 'La Portée, pp. 121–2.

15 Klinck, *Die lateinische Etymologie*, pp. 46–7. The *translatio vicinitatis* is in turn subdivided into seven possibilities: *per efficientiam, per affectum, per id quod continet, per id quod continetur, per abusionem, a parte totum, a toto pars*. A closer examination of the examples listed under these different headings reveals that pure analyses of the signifier are interwoven with analyses that belong to the process of metonymic thought.

16 Fontaine, *Isidore de Séville*, p. 671.

17 Lindsay (ed.), *Isidori*, XI, 1, 147.

18 Ibid., XI, 1, 97.

19 Ibid., XI, 1, 98. We will return to this localization (below, p. 138).

20 Ibid., XI, 1, 102.

21 Ibid., XI, 2, 19.

22 Ibid., XI, 2, 24 (translation modified).

23 For a fuller account, see W.D. Sharpe, 'Isidore of Seville: The Medical Writings', *Transactions of the American Philosophical Society*, new series, LIV–2 (1964), pp. 5–75.

24 Lindsay (ed.), *Isidori*, XI, 1, 140.

25 Rabanus Maurus, *De Universo*, XXII, 6; see J.-P. Migne, *Patrologiae Cursus Completus, Patres Latini* (Petit-Montrouge, Paris, 1852), columns, 173–4; moralization (column 175): 'Ad mulierem menstruatam non licet accedere nec cum ea commisceri, quia nec idolatriae paganorum nec haeresi haereticorum licet catholico homini communicare.'

26 Lindsay (ed.), *Isidori*, XI, 1, 137.

27 F. Redeker, 'Die *Anatomia Magistri Nicolai Physici* und ihr Verhältnis zur *Anatomia Cophonis* und *Ricardi*' (unpublished dissertation, University of Leipzig, 1917).

28 M. de Boüard, *Une nouvelle encyclopédie médiévale: le 'Compendium Philosophiae'* (E. de Boccard, Paris, 1936).

29 Jacopo Berengario da Carpi, *Commentaria cum Amplissimis Additionibus super Anatomia Mundini* (Hieronymus de Benedictis, Bologna, 1521), folio CCVIII verso and CCLIX recto; for an account of the life and work of Berengario da Carpi (1460–1530?), see Charles Coulston Gillispie (ed.), *Dictionary of Scientific Biography*, 16 vols (Charles Scribner's Sons, New York, 1970–80), vol. I, p. 619.

30 Jacques Despars, *Expositio supra Librum Canonis Avicenne* (Jean Trechsel, Lyon, 1498), book I, fen 1, doctrine 4, ch. 2. Jacques Despars is here relying on Vindicianus: 'The thin membrane, the veil formed of very fat veins, which covers and warms the belly or intestines, is called *epiplos* in Greek, and *omentum* or *mappa* in Latin): V. Rose, (ed.), *Theodori Prisciani Euporiston Libri III* (B.G. Teubner, Leipzig, 1894), p. 475. The differences between the two quotations are a perfect illustration of the growing complexity of anatomical vocabulary.

31 V.C. Bussemaker and C. Daremberg, *Oeuvres d'Oribase*, 6 vols (J.-B. Baillière, Paris, 1851–76), vol. III (1858), p. 352.

32 There is a long bibliography on this subject. It can be found in the article we are here drawing on: C. Maccagni, 'Frammento di un Codice di Medicina del Secolo XIV (Manoscritto N. 735, già Codice Roncioni N. 99) della Biblioteca Universitaria di Pisa', *Physis*, 11 (1969), pp. 311–78.

33 As well as in the article quoted above, a clear reproduciton of this diagram can be found in C. Singer, 'A Thirteenth-Century Drawing of the Anatomy of the Uterus and Adnexa', *Proceedings of the Royal Society of Medicine*, 9 (1916), pp. 44–5.

34 Miniatures reproduced in E. Wickersheimer, *Anatomies de Mundino dei Luzzi et de Guido de Vigevano* (E. Droz, Paris, 1926).

35 Ibid., p. 82.

36 P. de Koning (tr.), *Trois traités d'anatomie arabe* (E.J. Brill, Leyden, 1903), p. 387.

37 Ibid.

38 Our translation is made from the edition of the *Pantegni* (B. Trot and J. de Platea, Lyon, 1515), *Theorica*, III, 32. It must be noted that the chapters devoted to the genital organs in the *Pantegni* have sometimes been transcribed separately under the title *De Genitalibus Membris*. It is perhaps this separate transcription that Peter the Deacon calls the *De Genecia* of Constantine. An edition of these chapters can be found in: M. Green, 'The *De Genecia* attributed to Constantine the African', *Speculum*, 62 (1987), pp. 299–323.

39 Koning (tr.), *Trois traités*, p. 387.

40 *Pantegni*, loc. cit.

41 Koning (tr.), *Trois traités*, p. 389.

42 *Pantegni*, loc. cit.

43 While the actual text of the *Aphorisms*, probably based on the ancient translation from Greek into Latin, gives in book V the word *uterus*, the Galenic commentary, translated by Constantine, uses *vulva*, in aphorism 36, and *matrix* and *vulva* in aphorism 45; see the edition published by O. Scoti (Venice, 1523), fols 93 v–95 v.

44 Note that in Cicero, *natura* designates the sexual organs of men and women. It is applied to women in the *De Divinatione*: 'Parere quaedam matrona cupiens, dubitans essetne pregnans, visa est in quiete obsignatam habere naturam.' See R. Giomini (ed.), *M. Tulli Ciceronis Scripta quae Manserunt Omnia* (B.G. Teubner, Leipzig, 1975), fascicule 46, II, 70 (145).

Celsus apparently uses *naturale* and *naturalia* to designate both the external female organs and the entrance to the vagina. Discussing the extraction of a stone in the bladder, he writes: 'Sed virgini subjici digiti tamquam masculo, mulieri per naturale ejus debent. Tum virgini quidem sub ima sinisteriore ora; mulieri vero inter urinae iter, et os pubis incidendum est.' And, on the imperforate vagina: 'Earum [feminarum] naturalia nonnumquam, inter se glutinatis oris, concubitum non admittunt.' See C. Daremberg (ed.), *De Medicina*, (B.G. Teubner, Leipzig, 1859), VII, 26 and 28, pp. 312 and 316.

45 R. Radicchi, *La 'Gynaecia' di Moschione: Manuale per le Ostetriche e le Mamme del VI Sec. d. C. (Traduzione Italiana e Note con Testo Latino Tratto dai Codici e Ampio Glossario)* (Editrice Giardini, Pisa, 1970), p. 42. One may also read in Moschion the description of a hypertrophy of the clitoris, and its amputation: 'De inmoderata landica quam Graeci yos nymfin appellant.

Turpitudinis symptoma est grandis yos nymfe; quidam vero adseverant pulpam ipsam erigi similiter ut viris et quasi usum coitus quaerere. Curabis autem eam sic: supinam iactantes pedibus clusis myzo quod foris est et amplius esse videtur, tenere oportet et scalpello praecidere, deinde conpetenti diligentia vulnus ipsum curare' (ibid., p. 190).

46 See E. Bourdelle and C. Bressou, *Anatomie régionale des animaux domestiques*, 2nd edn (J.-B. Baillière, Paris, 1964), III, 'Le Porc', p. 43.

47 William of Conches, *Dialogus de Substantiis* (J. Rihelius, Strasbourg, 1567), pp. 241–2 (womb), p. 253 (stomach). Note that the *Dialogus* is frequently called *Dragmaticon Philosophiae*. The description of the villous interior of the stomach is also drawn from the *Pantegni, Theorica*, III, 25.

48 G.W. Corner, *Anatomical Texts of the Earlier Middle Ages*, Carnegie Institution of Washington Publication no. 364 (National Publishing Company, Washington, 1927).

49 Modern editions of the *Anatomia Cophonis*: S. de Renzi, *Collectio Salernitana* (Filiatre-Sebezio, Naples, 1854) vol. II (photographic reimpression, Bologna, n.d., pp. 388–90; I. Schwarz, *Die medizinischen Handschriften der kgl. Universitäts-bibliothek in Würzburg: Beschreibendes Verzeichnis mit literarischen Anmerkungen* (A. Stuber, Würzburg, 1907); Corner, *Anatomical Texts*, pp. 48–50. In a recent appraisal, M.H. Saffron opts for a date of composition immediately prior to the spread of Constantine's translations, around 1080–90: 'Salernitan anatomists', in C.C. Gillispie (ed.), *Dictionary of Scientific Biography* (1975), vol. XII, pp. 80–3.

50 Editions of the *Second Salernitan Demonstration*: Renzi, *Collectio Salernitana*, pp. 391–401; K.H. Benedict, 'Die Demonstratio Anatomica Corporis Animalis auf Grund einer Nachprüfung des Breslauer handschriftlichen Textes und eines Vergleiches mit einer Erfurter Handschrift neu herausgegeben' (unpublished dissertation, University of Leipzig, 1920). For an English translation, see Corner, *Anatomical Texts*, pp. 54–66.

51 Editions of these various texts: Redeker, 'Die *Anatomia Magistri Nicolai Physici*'; K. Sudhoff, 'Der *Micrologus*-Text der *Anatomia* Richards des Engländers', *Archiv für Geschichte der Medizin*, 19 (1927), pp. 209–39. For an English translation of the *Anatomia Magistri Nicolai Physici*, see Corner, *Anatomical Texts*, pp. 67–86.

52 Ibid., p. 50.

53 Not having been able to consult the edition by K. Benedict, we are using the one by Renzi, *Collectio Salernitana*, pp. 398–9, and the translation by Corner, *Anatomical Texts*, pp. 63–4.

54 See Koning (tr.), *Trois Traités*, p. 423.

55 Renzi, *Collectio Salernitano*, p. 399; Corner, *Anatomical Texts*, p. 64.

56 Redeker, 'Die *Anatomia*', pp. 56ff.; Corner, *Anatomical Texts*, pp. 84–5.

57 Isidore of Seville writes: '"Dubius", incertus, quasi duarum viarum', see Lindsay (ed.), *Isidori*, X, 77. As the *Synonyma* of Simon of Genoa (end of thirteenth century – beginning of fourteenth century) attest, the original meaning of the Greek *didumos* ('double') was known in the Middle Ages: '"Didimus" grece geminus vel gemelus' – Synonyma Simonis Genuensis (A. Zaroti, Milan, 1473). While the word *didumos* in Galen means 'testicle', the word's meaning fluctuates in the Middle Ages: for Nicholas, it refers to the sperm ducts, but for the fourteenth-century surgeon Henry of Mondeville, to the epididymis – he also retains a connotation of 'doubt', but he explains

it differently: 'For we must always fear its loosening and rupturing' (tr. I,
ch. 11), see E. Nicaise, *Chirurgie de maître Henri de Mondeville* (F. Alcan, Paris,
1893), p. 82.

58 Redeker, 'Die *Anatomia*', p. 52; Corner, *Anatomical Texts*, p. 75.

59 This idea was clearly expressed in the Hippocratic treatise *On Generation*, II,
2. See vol. XI of the edition and French translation by R. Joly (Les Belles-
Lettres, Paris, 1970), p. 45. The treatise states:

Those who have undergone an incision near their ear can have sexual
relations and ejaculate, but they emit little sperm, and it is weak and sterile.
This is because most of the sperm comes from the head, through the ears,
towards the spinal marrow; and this path, after the incision which has left
a scar, has hardened over. In children, the veins, which are narrow and
filled-up, prevent the sperm from moving along, and arousal does not take
place in the same way.

The same idea is encountered in the discussion of the Scythians in Hippocrates,
Airs, Waters, Places, tr. W.H.S. Jones (Loeb Classical Library, Cambridge,
MA and London, 1962), 22, which the medieval Latin version translated as
follows: 'Quibuscumque vene post auriculas inciduntur eis ultra prorsus ab
omni generatione orbantur'; G. Gundermann (ed.), (Bonn, 1911).

60 *Pantegni, Theorica*, II, 12.

61 Bibliothèque nationale, Latin MS 7027, ninth century, fol. 156. The form
quilin refers to the contemporary name given to the vene cava, namely *vena
chilis*. It is probably as a result of repeated transcriptions and deformations
that a specifically female vein was 'invented', whereas originally only a
branch of the vena cava was meant.

62 *Spurii Libri Galeno Ascripti* (Giunta, Venice, 1597), fol. 59 r.–60 v.; Italian
tr. V. Passalacqua, '*Microtegni seu De Spermate*', *Traduzione e Commento* (Istituto
di Storia della medicina dell'Università di Roma, Rome, 1958).

63 J. Willis (ed.), *Commentarii in Somnium Scipionis* (B.G. Teubner, Leipzig, 1970).
The text states, with special reference to the number seven: 'Hic denique est
numerus qui hominem concipi, formari, edi, vivere, ali ac per omnes aetatum
gradus tradi senectae atque omnino constare facit' (I, 6, 62). The parts of
the body are likewise divided into groups of seven: 'In aperto quoque septem
sunt corporis partes, caput, pectus, manus, pedesque et pudendum item quae
dividuntur non nisi septem compagibus iuncta sunt: ut in manibus est
umerus, brachium, cubitus, vola et digitorum modi terni' (I, 6, 77). This
number thus governs the entire body: 'Unde non immerito hic numerus,
totius fabricae dispensator et dominus, aegris quoque corporibus periculum
sanitatemve denuntiat' (I, 6, 81). On the origin of the theory of the seven
cells, see F. Kudlien, 'The Seven Cells of the Uterus: The Doctrine and its
Roots', *Bulletin of the History of Medicine*, 39 (1965), pp. 415–23.

64 Bourdelle and Bressou, 'Le porc', p. 43:

The uterus is bicornate. The horns are long, narrow, very convoluted, shaped
like the intestines, and linked behind in a short body which no exterior lign
of demarcation separates from the vagina. The uterine cavity is not spacious
and links directly with the vaginal cavity. It is covered by a wrinkled and

very hairy mucus membrane from which rise regular ligns of papillary projections around the cervix.

65 Avicenna, *Canon*, tr. Gerard of Cremona. (P. de Lavagnia, Milan, 1473), bk III, fen 20, 1, ch. 1.
66 Wickersheimer, *Anatomies de Mondino dei Luzzi et de Guido de Vigevano*, p. 24.
67 R. von Töply (ed.), *Anatomia Ricardi Anglici* (circa 1242–1252) ad fidem codicis ms. n. 1634 in Bibliotheca Palatina Vindobonensi (Vienna, 1902). For an English translation, with a refutation of the attribution of the text to Ricardus Anglicus, see Corner, *Anatomical Texts*, pp. 87–110. Corner places the composition of the *Anatomia Vivorum* around 1225.
68 See C. Ferckel, *Die Gynäkologie und ihre Quellen*, Alte Meister der Medizin und Naturkunde, V (C. Kuhn, Munich, 1912).
69 Töply (ed.), *Anatomia Vivorum*, p. 21.
70 '[Matrix] septem habet cellulas humana figura ad modum monetae impressas': William of Conches, *Dragmaticon Philosophiae*, (Strasbourg, 1567 edn), p. 24. This again refers to Macrobius's commentary on *The Dream of Scipio*: 'Verum semine semel intra formandi hominis monetam locato hoc primum artifex natura mollitur' (ed.), Willis, *Commentarii in Somnium Scipionis*, I, 6, 63.
71 Töply (ed.), *Anatomia Vivorum*, p. 21. This description is taken up by Henry of Mondeville, at the beginning of the fourteenth century: 'Furthermore, the cervix in its cavity, between its two openings, penetrates through numerous whorls and folds that fit into one another like the petals of a rose before it opens, or like the mouth of a purse closed by a cord, so that only urine may come out until the moment of delivery' (tr. I, ch. 9), see Nicaise, *Chirurgie de maître Henri de Mondeville*, p. 75.
72 Thomas of Cantimpré, *Liber de Natura Rerum*, ed. H. Boese (De Gruyter, Berlin and New York, 1973), I, LXI, p. 67. Likewise, Vincent of Beauvais says: 'Membrum vero marium coitui conveniens, id est virga, inter omnia alia membra crescit et minuitur sine aliqua subiecti lesione. Substantia enim eius talis est ut possit ei augmentum et diminutio convenire. Ideo namque creatio eius est ex nervo et cartilagine et extenditur a vento illo adveniente et augmentum quidem eius conveniens est coitui' – *Speculum Naturale* (D. Nicolini, Venice, 1591), bk 21, ch. 45, fol. 271 v.
73 Aristotle, *Parts of Animals*, tr. A.L. Peck, (Loeb Classical Library, Cambridge, MA, and London, 1961), 689a, p. 385. Aristotle's statement may be explained by observations carried out on animals, if necessary on the monkey and then improperly applied to man. Cf J. Ruffié, *Le Sexe et la mort* (Odile Jacob Seuil, Paris, 1986), p. 138: 'Finally, as in all placental mammals, the penis contains in the distal part a penile bone or *baculum*. As opposed to preceding groups, this bone exists only in a vestigial state in the great apes; it has completely disappeared in man.'
74 H. Stadler (ed.), *Albertus Magnus De Animalibus*, Beiträge zur Geschichte der Philosophie des Mittelalters, 15–16 (Aschendorff, Münster, 1916–20), bk I, § 467, p. 167.
75 Ibid., II, § 42, p. 241: 'Nemo autem arbitretur, quod quando dicimus osseam aut cartilaginosam esse virgam, intelligamus veram ossie aut cartilaginis essentiam, sed potius substantiam in duritie cartilagine et ossi secundum aliquid proportionatam.'
76 Avicenna, *Canon*, bk III, fen 20, tr. 1, ch. 1.

77 C. Thomasset (ed.), *Placides et Timéo ou Li secrés as philosophes*, Textes littéraires français (Droz, Paris and Geneva, 1980), §§ 228–31, pp. 100–2; C. Thomasset, *Une vision du monde à la fin du XIIIe siècle. Commentaire du Dialogue de Placides et Timéo* (Droz, Geneva, 1982), p. 120).

78 H. Stadler (ed.) *Albertus Magnus, De Animalibus*, bk IX, see also D. Jacquart and C. Thomasset, 'Albert le Grand et les problèmes de la sexualité', *History and Philosophy of the Life Sciences*, III (1981), pp. 86–7.

79 'De matrice nunc aliqua dicemus. Primo quare animalia coeunt? Respondetur secundum Avicennam super secundo de anima'; see L.R. Lind, *Problemata Varia Anatomica*, Humanistic Studies, 38 (Kansas University Press, Lawrence, 1968), p. 55.

80 On the practice of autopsies and the beginning of dissections at Bologna, a summary and bibliographical guide can be found in: N. Siraisi, *Taddeo Alderotti and his Pupils* (Princeton University Press, Princeton, 1981), pp. 110–14, 297–300; for an account of the life and works of Mondino de' Luzzi, see C.C. Gillispie (ed.), *Dictionary of Scientific Biography*, vol. IX (1974) pp. 467–9.

81 See P. Huard and M.D. Grmek, *Mille ans de chirurgie en Occident: V^e–XVe siècles* (R. Dacosta, Paris, 1966), pp. 30–32.

82 For an account of the life and works of Henry of Mondeville, see E. Wickersheimer, *Dictionnaire biographique des médecins en France au Moyen Age* (Geneva, Droz, 1979) (reprod. of 1936 edn.), pp. 282–3, and D. Jacquart, *Suppléments* (Geneva, 1979), pp. 117–8.

83 William of Saliceto, *Chirurgia* (B. Locatelli, Venice, 1502), bk IV, ch. 4.

84 Henry of Mondeville states merely: 'Avicenna indicates a third hole, through which involuntary emissions are supposed to take place' (Nicaise, *Chirurgie*, p. 81). And Mondino refers to dissection: 'Et ut melius anathomiam eius videas, debes separare ossa femoris et elevare virgam cum vesica et intestino recto vel sine illis et dividere virgam in longitudinem usque ad canalem eius et tunc in principio eius apparebunt tibi duo foramina': see Wickersheimer, *Anatomies*, p. 29.

85 Gentile da Foligno, *Expositiones in Librum Tertium Canonis Avicenne* (P. Mauter, Padua, 1477), bk III, fen 20, tr. 1, ch. 1.

86 Galen, *On the Usefulness of the Parts of the Body*, tr. from the Greek with an Introduction and Commentary by Margaret Tallmadge May, 2 vols (Cornell University Press, Ithaca, NY, 1968), vol. II, bk 14, ch. 11, p. 644; C.G. Kühn (ed.), *Claudii Galeni Opera Omnia* (G. Olms, Hildesheim), vol. IV (1964) (photographic reimpression of the Leipzig edn of 1822), p. 189.

87 Galen, *De l'utilité des parties du corps*, tr. C. Daremberg, *Oeuvres anatomiques, physiologiques et médicales de Galien*, vol. II (J.-B. Baillière, Paris, 1856), p. 118; H.J. von Schumann, *Sexualkunde und Sexualmedizin in der klassischen Antike* (E. Reinhardt Verlag, Munich, 1975), column 76.

88 Despars, *Expositio supra Librum Canonis Avicenne*, bk III, fen 20, tr. 1, ch. 1.

89 The translation from Arabic into Latin, known under the title *De Iuvamentis Membrorum*, stopped at book X; it thus left out the essence of Galenic teaching on the reproductive organs. The complete translation by Nicholas of Reggio, based on the Greek, was finished only in 1317, thus after the composition of the *Anatomia* of Mondino de' Luzzi. See also R.J. Durling, 'Corrigenda and Addenda to Diels' Galenica', *Traditio*, 23 (1967), p. 473; and 37 (1981), p. 380.

90 'Mulier quam anathomizavi anno preterito scilicet 1314 anno Christi de mense Ianuarii in maiorem in duplo hebebat matricem quam illa quam anathomizavi eodem anno de mense Martii . . . et propterea maior centies erat matrix porce quam anathomizavi 1316 quam unquam vidi in femina humani, potuit tum alia esse causa quia erat pregnans et in utero habebat 13 porcellos'; see Wickersheimer, *Anatomies*, p. 25.

91 Gentile da Foligno, *Expositiones*, bk III, fen 21, tr. 1, ch. 1.

92 Galen, *On the Usefulness*, II, bk XIV, ch. 8, p. 638; Kühn (ed.), *Opera Omnia*, p. 176.

93 Avicenna, *Canon*, bk I, fen 1, doctrine 5, section 5, ch. 3 and 5; bk III, fen 1, tr. 1, ch. 1.

94 Wickersheimer, *Anatomies*, p. 25.

95 Ibid., p. 73. For an account of the life and works of Guy of Vigevano, see Wickersheimer, *Dictionnaire biographique*, pp. 216–17 and Jacquart, *Supplément*, p. 97.

96 H. Stadler (ed.), *De Animalibus*, tr. 2, ch. 24, p. 164, § 458.

97 Nicaise, *Chirurgie*, bk I, ch. 9, p. 75.

98 Wickersheimer, *Anatomies*, p. 25.

99 Michael Savonarola, *Practica Maior* (Giunta, Venice, 1547), tr. VI, ch. 21.

100 Wickersheimer, *Anatomies*, p. 25.

101 Savonarola, *Practica Maior*, tr. VI, ch. 21. The source can be found in the *Pantegni, Theorica*, bk III, ch. 32: '[Natura feminea] deforis habet frustula de pellibusque vocantur badedera, que frustula sunt in feminis sicut preputia in masculis.'

102 E. Nicaise, *La Grande Chirurgie de Guy de Chauliac* (F. Alcan, Paris, 1890), p. 67. Middle English tr. in Margaret S. Ogden, *The Cyrurgie of Guy de Chauliac* (Oxford University Press, Oxford, 1971), p. 67. For an account of the life and works of Guy of Chauliac (died 1368), see Wickersheimer, *Dictionnaire biographique*, pp. 214–15, and Jacquart, *Supplément*, pp. 95–6.

103 Juvenal says of Messalina: 'Mox lenone suas iam dimittente puellas tristis abit, et quod potuit tamen ultima cellam claudit, adhuc ardens rigidae tentigine volvae et lassata viris necdum satiata recessit.'

The published French translation, which is deliberately not literal, does not catch the precise meaning: 'When the brothel-keeper dismisses the girls, she regretfully withdraws; all that she can do is close the last cell. Still burning with the *itch of her senses* all vibrant, she departs, weary of man, but not satiated.' See Juvenal, *Satires*, ed. and tr. P. De Labriolle and F. Villeneuve (Les Belles-Lettres, Paris, 1971), VI, lines 126–129, p. 64. (An English translation can be found in Juvenal and Persius, *The Satires*, tr. G.G. Ramsay, Loeb Classics [Harvard University Press and William Heinemann, Cambridge, MA and London, 1918], Juvenal, *Satires*, VI, lines 127–30, p. 93: 'And when at length the keeper dismissed the rest, she remained to the very last before closing her cell, and with passion raging hot within her went sorrowfully away: then exhausted but unsatisfied . . .').

104 As in the preceding quotations, we are using for ease of reference the translation by Nicaise, *Chirurgie*, tr. 1, ch. 9, p. 75. Note also the edition of the Latin text by J.L. Pagel (A. Hirschwald, Berlin, 1892) and of the medieval French text by A. Bos (Société des anciens textes français, Paris, 1897–8).

105 Galen, *On the Usefulness*, bk XV, ch. 3, pp. 660–1: 'As for the outgrowths

of skin at the ends of the two pudenda, in woman they [the labia majora and minora] were formed for the sake of ornament and are set in front as a covering to keep the uteri from being chilled; in man it was impossible not to have them at all... and besides, they [the prepuce] serve as an ornament. The part called *nympha* [the clitoris] gives the same sort of protection to the uteri that the uvula gives to the pharynx; for it covers the orifice of their neck by coming down into the female pudendum and keeps it from being chilled'; Kühn (ed.), *Opera Omnia*, p. 223.

106 We have found nothing in the *Liber ad Almansorem*, contrary to what Henry of Mondeville's reference would lead us to expect. He probably confused this source with the *Surgery* of Albucasis: 'Tentigo fortasse additur super rem naturalem. ... Caro autem orta est caro que nascitur in orificium matricis donec impleat ipsum et fortasse egreditur ad exteriora secundum similitudinem caude' (B. Locatellus, Venice, 1500), ch. 71, fol. 23 v. Likewise, Avicenna mentions the clitoris only in the context of pathology (*Canon*, bk III, fol. 21, tr. 4, ch. 23).

107 For *badedera*, see above, p. 24. Gerard of Cremona uses *tentigo* in his translation of the *Surgery* of Albucasis and *batharum* in his translation of the *Canon* of Avicenna.

108 *Synonyma Simonis Genuensis*.

109 'Se comprimunt crura plicando unum crus super alterum, ut sic una partium vulve scalpat aliam, quia ex hoc oritur delectatio et pollutio': Stadler (ed.), *De Animalibus*, IX, § 7, p. 676. See also Jacquart and Thomasset, 'Albert le Grand', p. 87.

110 Pietro d'Abano, *Conciliator* (O. Scoti, Venice, 1521).

111 G. Fallopio, *Observationes Anatomicae* (A. Birckmann, Cologne, 1562), p. 300. For an account of the life and works of the famous anatomist, see *Dictionary of Scientific Biography*, vol. IV (1971), pp. 519–21.

112 'Ego autem vidi mulierem libidinosam que mihi ore suo dixit, quod similiter se vento exhibuit et multum delectabatur in conceptu venti intra uterum suum': Stadler (ed.), *De Animalibus*, VI, § 118, p. 492. This refers to the example of the mares quoted by Avicenna: 'Et equa, quando appetit, ventrem suum offert vento et delectatur in eo quod penetrat ipsam ventus et ex eo quod generatur in ventre suo ventus' – *De Animalibus* (B. Veneti, Venice, 1495), VII, ch. 1, fol. 36. For the whole tradition relating to stories of impregnation by the wind, see C. Zirkle, 'Animals Impregnated by the Wind', *Isis*, 25 (1936), pp. 95–117.

Chapter 2 *Physiology or the Stages of Purification*

1 This diagram is reproduced and discussed in, for example, Sharpe, 'Isidore of Seville: The Medical Writings', p. 24.

2 For a general treatment of Galen's doctrine, see R.E. Siegel, *Galen's System of Physiology and Medicine* (S. Karger, Basel and New York, 1968).

3 Ibid., p. 188.

4 See M. Ullmann, *Islamic Medicine* (Edinburgh University Press, Edinburgh, 1978), pp. 62–3.

5 Galen, *On the Usefulness*, bk XIV, ch. 10, pp. 641–2.

6 Ibid., bk XIV, ch. 7, pp. 634, 635–6.

7 See H. Rouvière, *Anatomie humaine, descriptive, topographique et fonctionnelle, onzième édition révisée et augmentée par A. Delmas* (Masson et Cie, Paris, 1974), vol. III, pp. 223–4.

8 Galen, *On the Usefulness*, bk XIV, ch. 7, p. 636.

9 This question is raised for instance in the *Conciliator* of Pietro d' Abano: *Differentia* 28 (O. Scotus, Venice, 1521), fol. 39. The conclusion reached is, of course: 'Masculus universaliter calidior et siccior est femella.' In the course of the discussion, one finds the argument traditionally put forward in favour of a supposed greater heat in females; washerwomen do not suffer from the cold, when they are washing sheets, even though their bare feet dip into freezing water. This argument is refuted in two ways: first, woman has an imperfect sense of touch which makes her less sensitive than man; secondly, like aquatic animals, she suffers less from the cold because she is cold in nature. With this last reason, the image of the serpent, which we shall be meeting again when we discuss menstruation, is already looming.

10 Lindsay (ed.), *Isidori*, XI, 1, 77.

11 'You have to know that the absorbed semen is like milk during the first six days, and that in the next nine days, blood appears' – S. de Renzi, *Flos Medicinae Scholae Salerni* (Filiastre-Sebezio, Naples, 1859), p. 51.

12 E. Lesky, 'Die Sametheorie in der hippokratischen Schriftensammlung', *Festschrift Max Neuburger* (W. Maudrich, Vienna, 1948), pp. 302–7; 'Die Zeugungs- und Vererbungslehre der Antike und ihre Nachwirkungen', *Abhandlungen der Wissenschaften und der Literatur in Mainz*, 19 (1950), pp. 1227–424. The works of Erna Lesky are examined by R. Hippeli and G. Keil, *Zehn Moden Menschenwerdung* (Basotherm GmbH, Biberach an der Riss, 1981).

13 *Airs, Waters, Places*, 22; H. Diller (ed.), *Corpus Medicorum Graecorum*, I, 1, 2 (Akademie-Verlag, Berlin, 1970), and with a translation into English (cited here) in Hippocrates, *Works*, 4 vols (Loeb Classical Library, Cambridge, MA, and London, 1923–31), vol. IV, tr. W.H.S. Jones, pp. 127–31. See also ch. 1, note 59 above.

14 *Timaeus*, tr. R.G. Bury (Loeb Classical Library, Cambridge, MA, and London, 1929), 91a and 91b, p. 249.

15 Hippocrates, *De la génération*, ed. and tr. Joly, VIII, 1, 480, p. 49.

16 'Semen . . . est enim liquor ex cibi et corporis decoctione factus, ac diffusus per venas atque medullas, qui inde desudatus in modum sentinae concrescit in renibus eiectusque per coitum' (*Etymologiae*, ed. Lindsay, XI, 1, 139).

17 *Dialogus de Substantiis*, pp. 236–7.

18 *Generation of Animals*, tr. A.L. Peck (Loeb Classical Library, Cambridge, MA, and London, 1943), 725a, p. 81.

19 See Siegel, *Galen's System*, pp. 224–30.

20 Avicenna, *Canon*, tr. Gerard of Cremona, bk I, fen I, doctrine 4, ch. 1; and bk III, fen 20, tr. 1, ch. 3.

21 The commentator supposes that the sperm comes from the fourth digestion at its very beginning, for if it were the final residue, it would be too thick to flow as far as the testicles.

22 *Differentia*, 34 ('Quod sperma a toto corpore sive ab omnibus decidatur membris ostenditum') (O. Scoti, Venice, 1521), fol. 50 r.

23 XV, Q14 'Utrum magis derivetur sperma ab una parte quam ab alia',

E. Filthaut (ed.), *Alberti Magni Opera Omnia*, (Aschendorff, Münster, 1955), vol. XII.

24 'Why do both the eyes and the buttocks sink in those who indulge excessively in sex, though the latter are near to and the former far from the sex organs?', Pseudo-Aristotle, *Problems*, I-XXI, tr. W.S. Hett (Loeb Classical Library, Cambridge, MA, and London, 1961), 2, pp. 108–9.

25 *Generation of Animals*, 747a, pp. 247 and 249.

26 Lind, *Problemata Varia Anatomica*, p. 56.

27 Vincent of Beauvais, *Speculum Naturale*, XXI, 52 (A. Rusch, Strasbourg, 1478).

28 For a discussion of this problem, see M.A. Hewson, *Giles of Rome and the Medieval Theory of Conception* (The Athlone Press, London, 1975), pp. 21–134.

29 Several of the condemned propositions deal with this problem; for example, 'Quod omnium formarum causa effectiva immediata est orbis' and 'Quod, sicut ex materia non potest aliquid fieri sine agente, ita nec ex agente potest aliquid fieri sine materia'; see R. Hissette, *Enquête sur les 219 articles condamnés à Paris le 7 mars 1277* (Publications Universitaires, Louvain, 1977), articles 81 and 108.

30 See D.E. Sharpe, 'The 1277 Condemnation by Kilwardby', *New Scholasticism*, 8 (1934), pp. 308–18.

31 On this subject we are relying on G. Durand, 'Anthropologie sexuelle et mariage chez saint Thomas d'Aquin' (unpublished dissertation for the Faculty of Theology of the University of Lyon, 1966–7, pp. 91ff).

32 Ibid., p. 90.

33 See Hewson, *Giles of Rome*. In a review of this work by Michael R. McVaugh, the date of 1276 proposed by Hewson is disputed and replaced by the years 1285–90: *Speculum*, 52 (1977), pp. 987–9.

34 'Unde et philosophus in XVI de animalibus dicit quod corpus spermatis, cum quo exit spiritus, quae est virtus principii animae, est separata a corpore et est res divina, quae omnia sic sunt intelligenda, quod quia talia facit ista virtus, quae non videtur facere corpus; ideo necesse est, quod agat in virtute alicuius principii separati, et ex hoc mercatur dici res divina, modus itaque agendi istius virtutis, qui sic agit modo alto, sufficienter arguit quod nullo modo fiat pars fetus nec se habet ut materia' – Giles of Rome, *De Formatione Corporis Humani in Utero* (J. Pentius de Leuco, Venice, 1523), p. 55.

35 *Giles of Rome*.

36 The reference is to a pun on Aphrodite, a name supposedly derived from *aphros* (foam); Aristotle, *Generation of Animals*, 736a, p. 165.

37 Nemesius of Emesa, *De Natura Hominis*, tr. Burgundio of Pisa, ed. G. Verbeke and J.R. Moncho (Brill, Leyden, 1975), p. xxvii.

38 *Dialogus de Substantiis*, p. 237.

39 J. Filliozat, *The Classical Doctrine of Indian Medicine: its Origins and its Greek Parallels*, tr. Dev Raj Chanana (Munshiram Manoharlal, Delhi, 1964), pp. 165–6.

40 See W. Gerlach, 'Das Problem des "weiblichen Samens" in der antiken und mittelalterlichen Medizin', *Sudhoffs Archiv*, 30 (1938), p. 178.

41 *Generation of Animals*, 727a, p. 97.

42 Ibid., 727b–8a, p. 101.

43 *De la génération*, IV, I, pp. 46–7.

44 *On the Usefulness*, II, bk XIV, ch. 11, p. 643.

45 Koning (tr.) *Trois traités*, p. 397.
46 Ibid., p. 772.
47 The English translation is taken from M.C. Seymour, (general ed.) *On the Properties of Things: John Trevisa's Translation of Bartholomaeus Anglicus 'De Proprietatibus Rerum': A Critical Text*, (Clarendon Press, Oxford, 1975), vol. I, p. 294, lines 17–33.
48 William of Conches, *Dragmaticon Philosophiae*, edn cited, p. 240.
49 Ibid., p. 241.
50 See *Placides et Timéo*, ed. Thomasset, pp. 256–61.
51 B. Lawn (ed.), *The Prose Salernitan Questions* (Oxford University Press, Oxford, 1979) B 46, p. 22.
52 Thomas of Cantimpré, *Liber de Natura Rerum*, ed. Boese, bk I, p. 72.
53 We are again drawing on Hewson's *Giles of Rome*, pp. 67–94.
54 'Multotiens quidem expertum est quod aliquis von volens violare virginem fecundat eam, quod hoc modo contingit: sperma fusum extra matricem caliditate matricis attractum fecundavit eam' (Giles of Rome, *De Formatione*, p. 40).
55 Averroes, *Colliget*, II, c. *De Iuvamentis Membrorum Virtutis Generative* (J. and G. de Forli, Venice, 1490), fol. 10 r.
56 See Hewson, *Giles of Rome*, pp. 89–90.
57 Albert the Great, *De Animalibus*, bk XV, tr. 2, ch. 6; Stadler (ed.), *Albertus Magnus De Animalibus*, pp. 1038–9, §§ 109–10; see also Jacquart and Thomasset, 'Albert le Grand', p. 84.
58 Albert the Great, *De Animalibus*, bk XI, tr. 1, ch. 1 and ch. 3, Stadler (ed.), *Albertus Magnus De Animalibus*, pp. 732–3,§ 6, and p. 741, § 28. The remark 'after a woman's period' suggests the intuitive knowledge that there is a periodicity in fecundity. This is also suggested by a text in Old French: 'All the times that the man lies with the woman do not lead to begetting; but certain times established by nature. This is either at the beginning of the purgation of the woman, which she has by nature every month, or else at the end of the purgation' (MS Paris, Bibliothèque nationale, new Latin acquisition 693, fol. 172 v.–173 v.). See also M. Faribault, 'Le réceptaire de Jean Sauvage', (unpublished dissertation for a 'thèse 3ᶜ cycle', University of Paris IV-Sorbonne, 1980, p. 257).
59 Filthaut (ed.), *Alberti Magni Opera Omnia*, vol. XXVIII (1951) p. 164.
60 Ibid., vol. XII (1955), p. 217.
61 A chapter of Avicenna's *Canon* is devoted to discharges from the uterus (bk III, fen 21, tr. 3, ch. 23); in it, the distinction between pathological and functional cases of leucorrhoea is suggested. The former are ascribed to an excess of humoral superfluities, the latter, compared to ejaculations in the male, are identified with sperm, as in Albert the Great.
62 Albert the Great, *De Animalibus*, bk XV, tr. 2, ch. 11, Stadler (ed.), *Albertus Magnus De Animalibus*, p. 1057, § 145.
63 See Jacquart and Thomasset, 'Albert le Grand', pp. 79–80.
64 *Conciliator: Differentia*, XXVII, fols 53–4.
65 *Practica Maior*, tr. VI, ch. 21, rubric 22.
66 See J.T. Noonan, *Contraception: A History of its Treatment by the Catholic Theologians and Canonists* (Harvard University Press, Cambridge, MA, 1966), p. 176.
67 René Descartes, 'La Description du corps humain et de toutes ses fonctions',

in C. Adam and P. Tannery, (eds.) *Oeuvres de Descartes*, new edn, (J. Vrin, Paris, 1974), vol. XI, p. 253. English translation from 'Description of the Human Body and of all its Functions', in *The Philosophical Writings of Descartes*, tr. John Cottingham, Robert Stoothoff and Dugald Murdoch, 2 vols (Cambridge University Press, Cambridge, 1985), vol. II, part 4, § 253, pp. 321–2.

68 Lindsay (ed.), *Isidori*, XI, 1, 140.

69 *Trotulae de Mulierum Passionibus ante, in et post Partum* (J. Schott, Strasbourg, 1564), p. 3.

70 C. Singer, 'A Thirteenth-Century Drawing of the Anatomy of the Uterus and Adnexa, *Proceedings of the Royal Society of Medicine*, 9 (1916), p. 46. On this diagram, see also above, p. 17.

71 Koning, *Trois traités*, p. 399.

72 Ibid., pp. 403–5.

73 Avicenna, *Canon*, bk I, fen 3, doctrine 1, ch. 2.

74 See Lawn, *The Prose Salernitan Questions*, B 300, p. 142.

75 *Generation of Animals*, 727a, p. 95.

76 See Lawn, *The Prose Salernitan Questions*, pp. 42 and 222.

77 Thomasset, *Commentaire du Dialogue*, pp. 138–9, §§ 300–1.

78 Albert the Great relates that poor women, who work hard and have harsh lives, do not have menstrual periods because what they eat is hardly enough to keep them alive (*Quaestiones super De Animalibus*, ed. Filthaut, IX, Q 5). He bases this opinion notably on Avicenna who indicates lack of food as being among the causes of 'scarcity of menses' (*Canon*, bk III, fen 21, tr. 3, ch. 25).

79 From the abundant literature devoted to this topic, we will quote: P.C. Racamier, 'Mythologie de la grossesse et de la menstruation', *L'Evolution psychiatrique*, (1955), pp. 285–97. Here is the analysis of different taboos:

Such are the principal unconscious themes that anthropological study reveals in the experience of menstruation. They tally with all the instinctual stages that psychoanalysis has described, and can be deduced from the following facts, which we will draw to the reader's attention: menses are a *discharge* (which can be excremental) of *blood* (which can be that of a wound and an attack) from the *genitalia* (definitively deprived of a phallus, perhaps as a result of incestuous sexual acts and criminal masturbation) of a *woman* (erotically stimulated) capable of *giving birth* (but who is not pregnant, and may be afraid that she cannot conceive). All the causes for joy, or else anguish and bitterness, that menstruation may thus contain, are easily understandable. During her period, the woman imagines herself – and is imagined by those around her, particularly men – to be a guilty, dirty, dangerously seductive person, and one who is above all to be *feared*.

These features are of course all found in the Middle Ages, in the very heart of scientific discourse.

80 Vincent of Beauvais, *Speculum Naturale*, XXXI, 24.

81 Several Salernitan questions deal with the dangers of the basilisk, previously discussed by Pliny (*Natural History*, VIII, 33, and XXIX, 19). See Lawn, *The Prose Salernitan Questions*, B 05–B 08, pp. 49–51, n° 30, p. 298. Isidore of Seville also mentions this animal (see Lindsay (ed.), *Isidori*, XII, 6).

82 *The Problemata Varia Anatomica* repeat in the same terms the idea expressed in the *De Secretis Mulierum*:

> Ad secundum respondetur quia basiliscus est animal venenosum et per oculos emittit humores venenos qui exeuntes multiplicantur usque ad rem visam et usque ad oculos hominis ubi inveniunt pellem et poros subtiles qui humores scilicet intrantes corpora inficiunt, et haec est causa quare basiliscus inspiciens speculum seu clipeum ex bitumine contextum aut aliam rem firmam interficit se ipsum, et quia ab ista se polita isti humores fumosi reflectuntur et repercutiunt usque ad bisiliscum ex qua repercussione inficitur et interficitur. Et sic similiter dicendum de muliere menstruosa. [Lindsay, (ed.) p. 20.]

83 *Quaestiones super De Animalibus*, ed. Filthaut, IX, Q 9; Albert the Great brings up this idea again in his commentary on the *De Anima*, ed. C. Stroick, *Alberti Magni Opera Omnia*, vol. VII, (1968), III, 1.

84 Aristotle, *Parva naturalia*, tr. W.S. Hett, Loeb Classics (Harvard University Press and William Heinemann, Cambridge, MA, and London, 1957), 459b, p. 356. The role played by the air is there clearly demonstrated; C. Gaignebet, 'Véronique ou l'image vraie', *Anagrom*, 7–8 (1976), pp. 45–70.

85 These questions have been discussed by C. Thomasset, 'La Femme au Moyen Age. Les Composantes fondamentales de sa représentation: immunité-impunité', *Ornicar*, 22–23 (1981) pp. 223–38.

86 *Les Admirables secrets de magie du Grand Albert et du petit Albert* (Paris, n.d.).

87 MS Paris, Bibliothèque nationale, Latin 7148, fol. 2 r. 9 v.

88 See a translation of the *Secrets of Women* quoted by Gaignebet, 'Véronique ou l'image vraie', p. 60.

89 *Les Secrés des dames*, ed. E. Rauneyre (Paris 1880), pp. 65–6.

90 See Durand, *Anthropologie sexuelle*, pp. 83–4.

91 Bk I, fen 1, doctrine 5, ch. 1.

92 Avenzoar, *Theisir*, tr. John of Capua, tr. 3, ch. 4 (O. Scoti, Venice, 1530), fol. 40 r.

93 *De Coitu* MS Paris, Bibliothèque nationale, Latin 16195, fol. 23 v–fol. 25 r., end of thirteenth century.

94 Bk IX, tr. 1, ch. 7, Stadler (ed.) *Albertus Magnus De Animalibus*, p. 700, § 66.

95 Bk III, fen 20, tr. 1, ch. 3.

96 See A. Rousselle, *Porneia: De la maîtrise du corps à la privation sensorielle, II^c–IV^c siècle de l'ère chrétienne*, Les Chemins de L'Histoire (Presses Universitaires Françaises, paris, 1983), p. 31.

97 *On the Usefulness*, II, Bk XIV, pp. 640–1.

98 The biological purpose of pleasure was reaffirmed in the *Viaticum* of Constantine the African (Lyon edn, 1510, Bk VI):

> Deus ad animalia genera existenda creavit membra unde essent procreanda. Quibus proprie virtutem indidit naturalem ex qua multum delectari possent et operari in ipsis membris animalibus et amabilem fecit cum maxima concupiscentia et desideriis et in coytu admirabilem dedit delectationem et inseparabilem, ut genus eorum reparetur ne forte abhominato coytu ab animalibus generatio ammitteretur.

99 *Super Ethica, bib. III, Lectio XIII*, ed. W. Kübel, *Alberti Magni Opera Omnia*, (1968–72), pp. 206–7.

100 Thomas Aquinas says the same thing: 'The excessive emission of semen is contrary to the good of nature, which is the preservation of the species'; see J.-L. Flandrin, *Le Sexe et l'Occident* (Seuil, Paris, 1981), p. 117.

101 Note that most theologians condemned these marks of affection, claiming that they risked leading to an involuntary emission; Thomas Sanchez authorizes 'the customary embraces, kisses and caresses between husband and wife, so that they may demonstrate and reinforce their mutual love' (see Flandrin, ibid., p. 107).

102 See Lawn, *The Prose Salernitan Questions*, B 7, p. 4.

103 B. Trot and J. De Platea edn, VI, 17.

104 Albert the Great, *De Animalibus* Bk X, tr. 1, ch. 1; Stadler (ed.), *Albertus Magnus de Animalibus*, p. 730, § 1. Juvenal's formula (*Satires*, VI, 130), is repeated by, among others, Vincent of Beauvais, *Speculum Naturale*, Bk 31, ch. 5.

105 Albert the Great, *Quaestiones super De Animalibus*, ed. Filthaut, V., Q 4 and 6. The idea of a female pleasure quantitatively greater, but weaker in intensity, than that of the male had already been expressed by Peter of Spain, cf. M.F. Wack, 'The Measure of Pleasure: Peter of Spain on Men, Women, and Lovesickness', *Viator*, 17 (1986), pp. 172–96. On the basis of similar data, Hildegard of Bingen (twelfth century) reached different conclusions, as Joan Cadden points out:

Woman is colder and moister, but it is these qualities that make her fertile: she is more spacious and her passions milder, and these characteristics allow her to conceive, carry, and give birth to children. . . . Since her desire is less violent, a woman is more able to contain herself – because of the moistness 'where the pleasure burns' and because of either fear or shame. [J. Cadden, 'It Takes all Kinds: Sexuality and Gender Differences in Hildegard of Bingen's Book of Compound Medicine', *Traditio*, 40 (1984), p. 159]

Hildegard's interpretation is altogether exceptional in the medieval context.

106 'Queritur cur cum cetera animalia post conceptionem a coitu cessant mulieres tunc libentius coeunt?' , ed. Lawn, *The Prose Salernitan Questions*, B 23, pp. 13–14. This type of question, frequently asked by different writers, should clearly be seen in the context of the religious prescriptions that govern sexual intercourse during pregnancy; on these prescriptions, see J.-L. Flandrin, *Un temps pour embrasser. Aux origines de la morale sexuelle occidentale (VIe–XIe siècle)* (Seuil, Paris, 1983).

107 Albert the Great, *Quaestiones super De Animalibus*, ed. Filthaut, I, Q 13.

108 'Queritur quare tanta delectatio sit in coitu?', Lawn, *The Prose Salernitan Questions*, B 16, pp. 10–11.

109 K. Garbers, *Ishâq ibn 'Imrân, 'Maqâla fî l-mâlikhûliyâ' und Constantini Africani Libri Duo de Melancholia* (Helmut Buske Verlag, Hamburg, 1977), pp. 99–100 and 185).

110 Avicenna, *Canon*, Bk III, fen 20, tr. 1, ch. 5.

111 For the medical and philosophical antecedents, both classical and Arabic, of this doctrine, see E.R. Harvey, *The Inward Wits: Psychological Theory in the Middle Ages and Renaissance* (Warburg Institute, London, 1975).

112 '*Epistola de Amore qui Dicitur Heroicus*', in Arnaldi Opera (F. Fradin, Lyon,

1504). Michael R. McVaugh is preparing a critical edition of this treatise for the *Arnaldi de Villanova Opera Medica Omnia* (Granada and Barcelona). We have analysed the shift from *erôs* to heroic love, and the courtly background to this diseased form of love, in: D. Jacquart and C. Thomasset, 'L'Amour "héroïque" à travers le traité d'Arnaud de Villeneuve', in J. Céard, (ed.) *La Folie et le corps* (Ecole normale supérieure, Paris, 1985), pp. 143–58. A translation made *c*.1100 – and thus later than Constantine's *Viaticum* – of the chapter on passionate love in the *Zâd al-musâfir* of ibn al-Jazzâr already uses the adjective *heroicus*, cf. M.F. Wack, 'The *Liber de Heros Morbo* of Johannes Afflacius and its Implications for Medieval Love Conventions', *Traditio*, 62 (1987), pp. 324–44.

113 See D. Jacquart, 'La Maladie et le remède d'amour dans quelques écrits médicaux du Moyen Age', *Actes du Colloque 'Amour, mariage et transgressions au Moyen Age'* (Kümmerle Verlag, Göppingen, 1984), pp. 93–101. Obscenity can be found, for instance, in a recipe of the *Thesaurus Pauperum* attributed to Peter of Spain, in *Obras Médicas de Petro Hispano*, ed. M.H. Da Rocha-Pereira (University of Coimbra Press, Coimbra, 1973), ch. 30: 'Item si quis maleficiatus fuerit ad nimis amandum aliquem vel aliquam, merdam illius, quem diligit, recens, ponatur mane in subtellari dextro amantis et calciet se: quam cito fetorem sentiet, solvetur maleficium.'

114 *Arnaldi Opera*, (Lyon, 1504 edn) fol. 23 v.–26 v.

115 Lawn, *The Prose Salernitan Questions*, B 46, pp. 22–3: 'Spiritus infecti sperma illud, illud vero infectum prout materia fetus suum format materiarum, unde fetus non a generante formam contrahit sed ab illo cuius proles esse deberet.'

116 We cannot here cite the abundant bibliography concerning the philosophical and poetic context of heroic love, but we will indicate two fundamental studies: J.L. Lowes, 'The Loveres Maladye of Heroes', *Modern Philology*, 11 (1913–14), pp. 491–546; O. Bird, 'The Canzone d'Amore of Cavalcanti According to the Commentary of Dino del Garbo', *Medieval Studies*, 1 and 2 (1940–1), pp. 150–203 and 117–60. See also M.M. Fontaine, 'La lignée des commentaires à la chanson de Guido Cavalcanti "Donna me prega"', in Céard, (ed.) *La Folie et le corps*.

117 According to Thomas Aquinas, for example, love completely attracts the *intentio* of the soul and may go as far as the performing of acts against nature; see J. Simonet, 'Folie et notations psycho-pathologiques dans l'oeuvre de saint Thomas d'Aquin', *Nouvelle Histoire de la psychiatrie* (Privat, Toulouse, 1983), pp. 60–1.

Chapter 3   Medicine and the Art of Love

1 See M. Foucault, *The History of Sexuality*, tr. Robert Hurley (Pantheon, New York, 1978), vol. I; *An Introduction*, ch. 3, 'Scientia sexualis'.

2 See J.T. Noonan, *Contraception: A History of its Treatment by the Catholic Theologians and Canonists* (Harvard University Press, Cambridge, MA, 1966) – this work contains the essential bibliography on the subject; J.L. Flandrin, *Le sexe et l'Occident* (Seuil, Paris, 1981), pp. 101–35, and *Un temps pour embrasser: Aux origines de la morale sexuelle occidentale (VI^c–XI^c siècle)* (Seuil, Paris, 1983); P.J. Payer, *Sex and the Penitentials: The Development of a Sexual Code,*

550–1150 (University of Toronto Press, Toronto–Buffalo–London, 1984).
3 J.T. Noonan, *Contraception*, p. 158.
4 Ibid., pp. 163–4.
5 See for example L.W.H. Wasserschleben, *Die Büssordnungen der abendländischen Kirche* (Goneger, Halle, 1851), p. 237.
6 J.T. Noonan, *Contraception*, p. 174.
7 Several recipes of this type are indicated in E. Wickersheimer, *Manuscrits latins de médecine du haut Moyen Age dans les bibliothèques de France* (CNRS, Paris, 1966). As an example, we will cite a manuscript from the beginning of the ninth century (Paris, Bibliothèque nationale, Latin 11218, fol. 108 v.), which presents a recipe 'ut mulier non concipiat' and, a few lines later, another recipe 'si abortum facere volueris'. Countless examples have been found by historians of medicine; we cannot go into them here.
8 Thus, a fifteenth-century manuscript repeats some of the advice given in the *De Passionibus Mulierum* of Trotula, without discussing its purpose (Paris, Bibliothèque nationale, Latin 6988 A, fol. 149 v.). A long list of syrups, potions, powders, fumigations, pessaries and plasters is introduced by the sole title *Contra Impregnationem Mulierum*.
9 One cannot help but remember the 'herb' worn by the priest Pierre Clergue in his illicit relations with Béatrice de Planissoles:

When Pierre Clergue wanted to know me carnally, he used to to wear this herb wrapped up in a piece of linen, about an ounce long and wide, or about the size of the first joint of my little finger. And he had a long cord which he used to put round my neck while we made love; and this thing or herb at the end of the cord used to hang down between my breasts, as far as the opening of my stomach [*sic*]. [E. Le Roy Ladurie, *Montaillou: Cathars and Catholics in a French Village 1294–1324*, tr. Barbara Bray (Scolar Press, London, 1978), p. 173.]

10 The theological position of Albert the Great is especially apparent in his commentary on the *Sentences*, on the question *Quae penitentia debetur illis qui venena sterilitatis procurant* (*In IV Sententiarum*, XXXI, D, article 18, ed. A. Borgnet, pp. 249–50).
11 In the article 'Pyrus' (pear-tree), Albert the Great indicates: 'Qui autem magicis insudant, dicunt quod radix pyri, et praecipue stiptica et tarde maturi portata et ligata super mulierem, impedit conceptum: et similiter si mulier super se vel iuxta habuerit pyra, difficulter pariet' – *De Vegetalibus et Plantis*, VI, I, ed. A. Borgnet, (L. Vivès, Paris, 1891), vol. X, p. 201. The literary use of the pear-tree to describe adulterous relationships reaches a kind of perfection in the Latin comedy *Lydia*, composed around 1175: not only is it in a pear-tree that the woman deceives and mystifies her husband, but her lover's first name is *Pirrus*. The metaphorical word-play involving the pear also clearly underlies this. See G. Cohen, (ed.) *La 'Comédie' latine en France au XIIᵉ siècle*, 2 vols (Les Belles-Lettres, Paris, 1931), vol. I, pp. 214–46.
12 Da Rocha-Pereira, (ed.) *Obras Médicas de Pedro Hispano*, pp. 235–71.
13 At the end of his chapter on methods of contraception, Rhazes merely gives a list of several techniques for inducing an abortion and for further details on the topic refers to other parts of his work: 'His preterea omnibus que in capitulo de menstruorum provocatione ac de facili partu diximus utendum est' – *Liber ad Almansorem* (Milan, 1481), V, 72.

14 Da Rocha-Pereira, (ed.) *Obras Médicas*, p. 259.

15 Rhazes not only advises that the woman's legs be raised, a position which makes it easier for the semen to be taken in, but also recommends that the man withhold his sperm until the woman has had her emission (*Liber ad Almansorem*, ibid.).

16 We shall be returning to this topic.

17 The treatise *On the Nature of the Child* (XIII, I) cites the case of a woman singer who, because of her profession, could not get pregnant: 'And so I asked her to jump, kicking her heels up to her buttocks. She had already done it seven times when the sperm flowed to the ground with a noise: on seeing this, the singer gazed at it in astonishment' – see Joly (ed. and tr.) *De la Nature de l'Enfant*, p. 55.

18 Wickersheimer, *Dictionnaire biographique*, vol. I, p. 9.

19 John Nider (1380–1438); see Noonan, *Contraception*, p. 273.

20 *In IV Sententiarum*, XXXI, G, article 24, ed. A.Borgnet, p. 263. See below, pp. 135.

21 The carnal aspect of courtly love seems difficult to refute. On this topic, critics often refer to the treatise by Andreas Capellanus which we shall be examining below; see A. Denomy, 'Fin'Amors: The Pure Love of the Troubadours: Its Amorality and Possible Source', *Medieval Studies*, 8 (1945), pp. 139–207, and F. Schlösser, *Andreas Capellanus: seine Minnelehre und das christliche Weltbild um 1200* (H. Bouvier, Bonn, 1980), pp. 370–82. Note that the very existence of an ideal of courtly love has been challenged by some scholars: see J.F. Benton, 'An Historical View of Courtly Love', in F.X. Newman, (ed.) *The Meaning of Courtly Love* (SUNY Press, Albany, 1968), pp. 19–42.

22 R. Nelli, *L'érotique des troubadours* (Privat, Toulouse, 1963), pp. 199–204, and *Erotique et civilisations* (Weber, Paris, 1972), pp. 144–9. It seems that one has to accept a diversification of practices in the ritual way courtly love was conducted: the possibility we refer to does not close the discussion. Psychoanalytic approaches to courtly love show the complexity of the phenomenon: cf. H. Rey-Flaud, *La Névrose courtoise*, Bibliothèque des Analytica (Navarin, Editeur, Paris, 1983); J.-Ch. Huchet, *L'Amour discourtois – La 'Fin'Amors' chez les premiers troubadours*, Bibliothèque historique (Privat, Toulouse, 1987).

23 These are conditions that encourage games with the sexual ambiguity of the object of desire. The game concerns not only the transformations of the object, but also the transvestism of the author. The constitution of the scientific text placed under the name of Trotula is in this respect exemplary (cf. *infra*, p. 221, n. 77). We would point out that it was clearly stated in the Middle Ages that the character who established and legitimized knowledge concerning women was Hermaphrodite; cf. Thomasset, *Commentaire du dialogue*, p. 162.

24 See P. Dronke, *Medieval Latin and the Rise of European Love-Lyric*, 2 vols (Clarendon Press, Oxford, 1968), vol. I, ch. 2.

25 R.H. Van Gulik, *Sexual Life in Ancient China* (E.J. Brill, Leyden, 1961), p. 47.

26 J.T. Noonan, *Contraception*, p. 299. Noonan's opinion does seem difficult to support. Peter P.A. Biller ('Birth-Control in the West in the Thirteenth and Early Fourteenth Centuries', *Past and Present*, 94 (1982), pp. 3–26) suggests, on the basis of sources from various origins, that at the beginning of the fourteenth century, the practice of *coitus interruptus* spread among married

couples. One of the channels along which information travelled could have been the priests themselves: 'Could any person who read the descriptions of impregnation in thirteenth-century encyclopaedias (such as those of Thomas of Cantimpré and Vincent of Beauvais) and the account of Onan's sin in Genesis (with its literal, contraceptive gloss) not have had a clear notion of *coitus interruptus* as a contraceptive method?' (article cited, p. 20, n. 70).

27 R. Bossuat, (ed.) *Li Livres d'Amours de Drouart la Vache* (Champion, Paris, 1926), lines 47–51.

28 Ibid., lines 96–101.

29 Ibid., lines 7547–54. It is disturbing to realize that the majority of the manuscripts still preserved of the *De Amore* of Andreas Capellanus belonged to clerics. CF. B. Roy, 'A la recherche des lecteurs médiévaux du *De Amore* d'André le Chapelain', *Revue de l'Université d'Ottawa/University of Ottawa Quarterly*, 55 (1985), pp. 45–73. Bruno Roy concludes his article with these words: 'Who, in the Middle Ages, read the *De Amore* – and how? They were Churchmen, especially Germanic Churchmen, at the end of the Middle Ages – and, it seems, they read it with great pleasure.' We cannot, however, conclude from this that they all read it in the same way as did Drouart la Vache.

30 Bossuat, *Li Livres d'Amours*, lines 4083–98.

31 Ibid., lines 3570–6.

32 This is the fifth question: lines 5749–78; see also the ages of love, lines 491–504.

33 The authors have not been able to consult the edition by Walsh (1982), and use for the Latin text the editino that had hitherto been standard: E. Trojel, *Andrea Capellani Regii Francorum 'De Amore' Libri Tres* (Libraria Gadiana, Copenhagen, 1892), p. 279. Our translation does use Walsh's edition and translation: see *Andreas Capellanus on Love*, ed. with an English tr. by P.G. Walsh, Duckworth Classical, Medieval and Renaissance Editions (Duckworth, London, 1982): the above Latin quotation is on p. 256 of this edition. – *Tr.*

34 The *Livre d'Enanchet*, which discusses a certain number of historical and political facts, includes a translation of part of the work of Andreas Capellanus. It drew at least the essential lesson from it. See W. Fiebig, *Das 'Livre d'Enanchet' nach der einzigen Hs. 2585 der Wiener Nationalbibl.* (Jena and Leipzig, 1938), p. 57):

So that the woman need not grant the will of her lover. And on this subject note that there are five reasons, for which woman need not grant her lover's will. The first is: so that she may be seen sometimes to refuse what is asked of her. The second: for if she grant it straightaway, he will think that she has already learnt how to do such a thing. The third: that it is sweeter to the asker if he has had to wait a long time for what he wants. The fourth: while waiting for this moment, she may give somewhat. *The fifth: for fear of getting pregnant.* [Summary of Andreas Capellanus *loquitur nobilior nobiliori*, in *Andreas Capellanus on Love*, Walsh (ed.) pp. 158ff.]

35 Bossuat, (ed.) *Li Livres d'Amours*, lines 4439–40.

36 Ibid., line 6384.

37 Ibid., lines 5911–50.

38 Ibid., lines 4495–542 (*De l'amour as vilains*).

39 M. Lazar, *Amour courtois et 'fin amors' dans la littérature du XIIe siècle* (Klincksieck, Paris, 1964), p. 271.

40 Noonan, *Contraception*, p. 298. The only reference to this practice is made by Huguccio in the twelfth century. This is how he describes it:

> To render the conjugal debt to one's wife is nothing other than to make for her a plenty of one's body for the wifely matter. Hence one often renders the debt to his wife in such a way that he does not satisfy his pleasure, and conversely. Therefore, in the aforesaid case, I can so render the debt to the wife and wait in such a way until she satisfies her pleasure. Indeed, often in such cases a woman is accustomed to anticipate her husband, and when the pleasure of the wife in the carnal work is satisfied, I can, if I wish, withdraw, not satisfying my pleasure, free of all sin, and not emitting my seed of propagation [*Summa*, 2, 13].

> Huguccio's point of view aroused no reaction for a whole century. Was it a procedure that people preferred to ignore, even though the texts that we are using show that it was in fact known about? It will have been noted that according to Huguccio's account, the male does not emit semen. For family reasons, in the fourteenth century, Peter de Palude claims that there is no sin providing that neither the woman nor the man ejaculate. The Church bases the notion of sin on emission. How did this affect the behaviour of clerics and laity? We are restricted to mere hypothesis; it is probable that the way the act finished had to depend on the degree of control possessed by each individual. There was an ideal to be attained, but it must nonetheless be noted that the idea of a wasted emission is mentioned by Andreas Capellanus, in connection with the comparison between adolescents and mature men.

41 Bossuat, (ed.) *Li Livres d'Amours*, line 6270.

42 Tr. Walsh, *Andreas Capellanus on Love*, p. 101, Latin text, p. 100: Mulier ait: In amoris curiam facillimus est inventus ingressus, sed propter imminentes amantium poenas ibi est perseverare difficile, ex ea vero propter appetibiles actus amoris impossibilis deprehenditur exitus atque durissimus. Nam post verum amoris curiae ingressum nihil potest amans velle vel nolle, nisi quod mensa sibi proponat amoris, et quod alteri possit amanti placere. Ergo talis non est curia appetenda; eius namque loci est omnino fugiendus ingressus, cuius libere non patet egressus. Tartareae etenim talis potest locus curiae comparari; nam, quum Tartari porta cuilibet intrare moretur aperta volenti, nulla est post ingressum exeundi facultas.

43 Tr. Walsh, *Andreas Capellanus on Love*, p. 185.

44 See the 'Introduction' to the French translation by C. Buridant, *Traité de l'amour courtois* (Klincksieck, Paris, 1974). The antithesis between the *sapiens amator* and the *stultus amator* should, he said, be seen in the context of control of the physiological mechanisms. The author quotes Paul Zumthor, 'Notes en marge du *Traité de l'amour* d'André le Chapelain' (p. 31), who claims that '*Sapientia* can be understood as being derived from *sapere* in the sense of *scire* = knowledge acquired from the technique of love.' There really is a technique, which is something quite different from a strategy. And the art of loving is not 'a practical manual of worldly relationships', as Robert Bossuat thought. As for the exclusion of peasants from the practice of love, we have already discussed the topic. The cleric tends to show an arrogance in this area that

is by no means an isolated occurrence in history.

45 B. Bowden, 'The Art of Courtly Copulation', *Medievalia et Humanistica*, 9 (1979), pp. 67–85. The same type of reading is suggested by Bruno Roy, 'André le Chapelain ou l'obscénité rendue courtoise', in P. Ruhe and R. Dehrens (eds), *Coll. Würzburg 1984, Mittelalter Bilder aus neuer Perspektive, Diskussions Anstösse zu 'Amour courtois', Subjektivität in der Dichtung und Strategien des Erzählens*, (W. Fink, Munich, 1985), pp. 59–73.

46 See J. Boswell, *Christianity, Social Tolerance and Homosexuality* (University of Chicago Press, Chicago, 1980), pp. 253–4.

47 See W. Wetherbee, 'The Function of Poetry in the *De Planctu Naturae* of Alain de Lille', *Traditio*, 25 (1969), p. 101.

48 H. Silvestre, 'Du nouveau sur André le Chapelain', *Revue du Moyen Age latin*, 36 (1980), pp. 99–106.

49 'Why was honesty given to husbands and beauty to young girls?' – Serlo of Wilton (died *c*.1181); I. Oberg, (ed.) *Acta Universitatis Stockolmiensis, Studia latina*, vol. 14 (1965), p. 106, line 13.

50 See B. Roy, 'L'Humour érotique au XVᶜ siècle, in *L'Erotisme au Moyen Age* (Ed. de l'Aurore, Montreal, 1977), pp. 155–71.

51 See A. Denomy, 'The *De Amore* of Andreas Capellanus and the condemnation of 1277', *Mediaeval Studies*, 8 (1946), pp. 107–49.

52 Walsh (ed.), *Andreas Capellanus on Love*, p. 287.

53 *Etymologiae*, XII, 7, 63. Note too the etymology that Isidore gives for the word *aves* (birds): 'They are called *aves* because they have no pre-determined routes, but lose themselves in unmarked paths (*avia*)', ibid., XII, 7, 3. It is easy to see how the word could be used metaphorically. It is also noteworthy that the kite – *milvus* – to which Andreas Capellanus also alludes (pp. 63 and 69) is an animal that is *mollis* according to Isidore (XII, 7, 58). As for the meaning that Andreas Capellanus gives to ◆falcon' (*falco, falconis*), it is expressed not so much in the bird's behaviour – as is the case with the kite – but in its name: 'Si me igitur noveris a meis degenerare parentibus, non contumeliosa milvi appellatione vocandus reperior, sed honorabili falconis vocabulo nuncupandus exsisto' (Walsh (ed.), *Andreas Capellanus on Love*, p. 68).

54 C. Segre, *Li Bestiaires d' Amours di maistre Richart de Fournival et li response du Bestiaire* (Riccardo Ricciardi, Milan and Naples, 1957); Bartholomew the Englishman, *De Proprietatibus Rerum* (edn cited).

55 P.E. Pierrugues, *Glossarium Eroticum Linguae Latinae* (H. Barsdorf Verlag, Berlin, rev. edn 1908), p. 387.

56 Tr. Walsh (ed.), *Andreas Capellanus on Love*, p. 75; Latin text, p. 74:

Nam quod ultra cuiusque noscitur pervenire naturam, modica solet aura dissolvi et brevi momento durare. Nam inter lacertivas fertur aves nasci quandoquam quasdam, quae sua virtute vel ferocitate perdices capiunt; sed, quia istud ultra ipsarum noscitur pervenire naturam, fertur quod in eis nisi usque an annum ab earum computandum nativitate haec non possit durare ferocitas.

Other allusions to the partridge can be found on pp. 62, 68 and 122 (Latin text). On the first occurrence, the text has 'perdix vel fasianus'.

57 Bossuat, (ed.) *Li Livres d'Amours*, p. 87 (lines 3025–9).

58 Walsh, (ed.) *Andreas Capellanus* pp. 128–9.

59 See Wetherbee, 'The Function of Poetry', p. 108.

60 E. Faral, *Les Arts Poétiques du XIIe et du XIIIe siècle* (H. Champion, Paris, 1962), p. 325.

61 Matthew of Vendôme, *Ars Versificatoria*, in Faral, *Les arts poétiques*, pp. 106–93 (for the examples we cite, see pp. 145 and 134). 'Milo', in Cohen, (ed.) *La 'Comédie' latine*, pp. 167–77.

62 Bossuat, (ed.) *Li Livres d'Amours*, p. 167 (lines 5815–32); pp. 20–1 (lines 710–15).

63 P. Mandonnet, *Siger de Brabant et l'averroïsme latin au XIIIe siècle* (Freiburg-im-Breisgau, 1899), pp. ccxxv–ccxxvi, and clxxxiii–clxxxiv. Current research suggests that Fr Mandonnet's interpretation needs to be somewhat modified: J.F. Wippel, 'The Condemnations of 1270 and 1277 at Paris', *The Journal of Medieval and Renaissance Studies*, 7 (1977), pp. 169–201; R. Hissette, *Enquête sur les 219 articles condamnés à Paris le 7 mars 1277* (Publications Universitaires and Vander–Oyez, Louvain–Paris, 1977). Nonetheless, for our purposes, the fact remains that in a specific intellectual milieu, there existed a certain freedom of thought.

64 Jean de Meung's joking comment, declaring that it is no more indecent to call the testicles ballocks than to call them *relics*, and vice versa, has been noted by all the commentators. One could read his remark as a bold assertion of the arbitrary nature of the sign. The word in its different uses has one or more obscene meanings and must be capable of being associated with a whole set of metaphors (see lines 19639ff. and 21617ff. of the edition by Félix Lecoy). The sign can be motivated in the erotic language of the clerics and the coherence of the metaphors, even the development of a narrative of this type, could be governed by the associative network of the marginal language. But these are merely Isidorean suggestions. On these problems, see D. Poirion, 'Les Mots et les choses selon Jean de Meung', *Information littéraire*, 26 (1974), pp. 7–11.

65 See J. Batany, *Approches du 'Roman de la Rose'* (Bordas, Paris, 1973).

66 D. Poirion, *Le Roman de la Rose* (Hatier, Paris, 1973), p. 198.

67 See the essential work by P.Y. Badel, *Le Roman de la Rose au XIVe siècle. Etude de la réception de l'oeuvre* (Droz, Geneva, 1980). On the opinion of Martin le Franc, see R. Louis, *Le Roman de la Rose. Essai d'interprétation de l'allégorisme érotique* (H. Champion, Paris, 1974), p. 132.

68 See the edition by Félix Lecoy, lines 13245–315.

69 See Jean-Charles Payen, *La Rose et l'utopie* (Editions Sociales, Paris, 1976). The author devotes a chapter to the ex-courtesan and also mentions Villon's Belle Heaumière. One ought perhaps to consider the literary function and psychoanalytic meaning of the always aging courtesan.

70 We owe this information to Armand Llinarès who was kind enough to pass on to us the text of the lecture 'Raymond Lull, un fou d'amour' that he gave at the Necker Hospital on 13 May 1982. *Tree of Knowledge*, XVI, 8, 14.

71 See R. Lapesa Melgar, (ed.) *Diccionario Histórico de la Lengua Española* (Real Academia española, Madrid, 1976).

72 B. Mario Damiani, (ed.) *Celestina* (Ediciones Cátedra, Madrid, 1982). Our principal references are to pp. 74–7, 112, 162–3, 170, 193–9 of this edition.

73 Chaucer mentions this influence when his Merchant sees Constantine, whom he calls 'the cursed monk', as a pimp and purveyor of aphrodisiac potions (cf. M. Bassan, 'Chaucer's *Cursed Monk*, Constantinus Africanus', *Mediaeval Studies*, 24 (1962), pp. 127–40, and P. Delany, 'Constantinus Africanus and Chaucer's Merchant Tale', *Philological Quarterly*, 46 (1967), pp. 560–66).

Note that the *De Coitu* derives from a work by ibn al-Jazzâr. It is not known whether it is the translation of an independent treatise or of a sort of somewhat enlarged adaptation of a chapter of the *Viaticum*. E. Montero Cartelle, (ed.) *Constantini Liber de Coitu, El tratado de Andrologia de Constantino el Africano* (University of Santiago de Compostela, Santiago de Compostela, 1983).

74 This passage deals with ridding oneself of the obsession:

Hec etenim que imprimit tales formas vel iterum forme delectationem afferentes acquirunt ex omni delectabili sensibus obiecto, ex quorum delectabilem facierum numero consistit balneum temperatum, confabulatio dilectorum, intuitus pulcrorum ac delectabilium facierum et etiam quantum est ex arte coitus, precipue si cum iuvenibus et magis delectationi congruis exerceatur [Arnald of Villanova, *De Amore Heroico*, (F. Fradin, Lyon, 1504)].

The same idea can, of course, be found in Avicenna, *Canon*, bk III, fen 1, tr. 4, ch. 23.

75 In particular in the *Viaticum* (bk VI).

76 See M. Gorlin, *Maimonides 'On Sexual Intercourse fi l-jimâ'*, Medical Historical Studies of Medieval Jewish Medical Works, I (Rambash Publ., New York, 1961).

77 An examination of the manuscripts demonstrates that the three treatises placed under the name of Trotula were written by men, probably at Salerno, at the end of the twelfth century or the beginning of the thirteenth. Although this group of treatises cannot be attributed to her, it does seem to be attested that a woman doctor, named Trotula, practised at Salerno in the twelfth century. Cf. J.F. Benton, 'Trotula, Women's Problems, and the Professionalization of Medicine in the Middle Ages', *Bulletin of the History of Medicine*, 59 (1985), pp. 30–53. We shall continue to call our author Trotula, because it is under this name that the Middle Ages placed these treatises. For ease of reference, we have used the J. Schott edition (Strasbourg, 1564).

78 An amusing example can be found in a sermon of William Peraldus (thirteenth century); it also bears witness to the rivalry between doctors and priests. The preacher notices that while the woman accepts having her hair shaved without a murmur if it is on doctor's orders, she does not obey the priest who tells her not to wear a wig (MS Paris, Bibliothèque nationale, Latin 16472, fol. 209 v.; we are grateful to David d'Avray for this item of information).

79 Another woman fasts for a month and has herself bled,
   So doing in order that she might have a perfectly pale face.
   For the woman without pallor seems to resemble a peasant;
   Pallor is comely, it is the real colour of the woman in love, she says.
   And this woman even covers her lips with various foul colours:
   But why does she seek by artifice another colour?

   Roger of Caen (died 1095), *De Contemptu Mundi*, ed. Thomas Wright, *Anglo-Latin Satirical Poets and Epigrammatists of the Twelfth Century*, (Longman and Co., London, 1872), vol. II, p. 186.

80 See M. Ullmann, *Die Medizin im Islam* (Brill, Leyden and Cologne, 1970), pp. 193–8.

81 See Thomasset, *Commentaire du Dialogue*.

82 German edn and tr.: G. Haydar, 'Kitâb fî l-bâh wa-mâ yuḥtâgu ilaihi min tadbîr al-badan fî sti'mâlihi' des Qusṭâ ibn Lûqâ' (unpublished dissertation for the University of Erlangen–Nuremberg, Erlangen, 1973); and N.A. Barhoum, 'Das Buch über die Geschlechtlichkeit (Kitâb fî l-bâh) von Qusṭâ ibn Lûqâ' (unpublished dissertation for the same university, 1974). We are grateful to Ursula Weisser for informing us of these dissertations.

83 Kitâb nuzhat al-aṣḥâb fî mu' âsharat al-ahbâḥ fî 'ilm al-bâh, German edn and tr. K. Hallak (J. Hogl, Erlangen, 1973); F. Mansour, ibid., 1975, and T. Haddad, ibid., 1976).

84 German edn and tr. H.M. El-Haw, 'Risâla fimâ yahtâg ilaihi 'r-rigâl wan-nisâ' fî sti'mâl al-bâh mimmâ yadurr wayanfâ' des At-Tîfâsî, ibid., 1970.

85 'Rushd al-labîb ilâ mu'âsharat al-'ḥabîb' (Anleitung des Einsichtigen des Umgangs mit der geliebten Person) des ibn Falîta, German edn and tr.: chs 1–3 by G. Al-Bayati, ibid., 1976; ch. 4 by A. Husni-Pascha, ibid., 1975; ch. 6 by B. Al-Khouri, ibid., 1975; chs 12–14 by E. Sabbagh, ibid., 1973.

86 See M. Grignaschi, 'L'Origine et métamorphoses du 'Sirr al-'asrâr' (Secretum secretorum), Archives d'histoire doctrinale et littéraire du Moyen Age, 43 (1977), pp. 7–112; and 'La Diffusion du Secretum secretorum (Sirr al-'asrâr), ibid., 47 (1981), pp. 7–70; J. Monfrin, 'La Place du Secret des Secrets dans la littérature française médiévaie', in Pseudo-Aristotle: The 'Secret of Secrets', ed. W.F. Evans and Ch. Schmitt (Warburg Institute, London, 1982), pp. 73–113.

87 Thomasset (ed.), Placides et Timéo, p. 123, § 270.

88 Secreta Mulierum (Antwerp edn, 1538).

89 For a recent account of the position, see B. Kusche, 'Zur Secreta Mulierum–Forschung', Janus, 62 (1975), pp. 103–23.

90 Les Secrets des hommes et des femmes composés par le Grand Albert traduits du latin en français (no place, n.d.).

91 J. Delumeau, La Peur en Occident (Fayard, Paris, 1978), pp. 304–45.

92 Avicenna, Canon, bk III, fen 20, tr. 1, ch. 38.

93 Yaḥyâ ibn Mâsawaih (John Mesue), Le Livre des axiomes médicaux, ed. D. Jacquart and G. Troupeau (Droz, Geneva, 1980), p. 149, aphorism 43.

94 Despars, Expositio supra librum Canonis Avicenne edn cited, bk III, fen 20, tr. 1, ch. 29.

95 Avicenna, Canon, bk III, fen 21, tr. 1, ch. 9.

96 Gentile da Foligno glosses the long passage from Avicenna with the words: 'Ponit secundum canonem. Non sunt male, quia non sunt multum calide, quia tunc conveniet ludus permixtione vera, scilicet in quo emittatur semen' (Venice edn, 1520).

97 'Quarto precipit [Avicenna] quod vir actu coiens in muliere consideret horam in qua ipsa fortiter sibi iuncta adheret et ipsius oculi rubere incipiunt quasi scintilantes et anhelitus eius fit altus et frequens et in verbis suis balbutire incipit, tunc enim est in puncto emittendi sperma' (Lyon edn, 1498).

98 Lilium Medicine (G. Rouillum, Lyon, 1550), p. 633.

99 'Mas excitare foeminam debet ac sollicitare ad coitum, loquendo, osculando, amplectendo, mamillas contractando, tangendo pectinem et perinaeum totamque vulvam accipiendo in manus et nates percutiendo hoc fine atque proposito ut mulier appetat venerem. . .et cum mulier incipit loqui balbutiendo, tunc debent se commiscere' – Rosa Anglica (Augsburg, 1595), p. 555. In addition, John of Gaddesden does not hesitate to refer, a little further on, to 'bawdy people' (p. 556).

100 'Si virga erigatur tarde post primum coitum et secundo velis coire mane,
    ut forte mulier quae non emisit sperma compleat desiderium suum (quod
    ex eo scitur, quia ipsa post coitum applicat se viro, eumque amplectitur et
    osculatur, et postea manum ex abrupto pronit ad virgam et testiculos viri,
    ut videat an sit paratus ad pugnam) tunc consultum est coitum repetere et
    ante super dorsum et renes iacere; quoniam iste decubitus efficit erectionem
    virgae' – Gaddesden, *Rosa Anglica*, p. 557.
101 Ibid.,   p. 557.
102 *Practica Maior*, tr. 6, ch. 21, rubric 23 (Giunta, Venice edn 1547).
103 *Expositio supra Librum Canonis Avicenne*, bk III, fen 20, tr. 1, ch. 38.
104 'Tertio debet vir mulierem tangere ut circa mamillas et leviter et specialiter
    capita mamillarum oscula nungere' (*Practica Maior*, loc. cit.).
105 Ibid., rubric 28, ch. 32.
106 See above, p. 46.
107 *Canon*, bk III, fen 20, tr. 1, ch. 6.
108 *Lilium Medicine*, p. 602.
109 *Practica Maior*, tr. 6, ch. 20, rubric 28.
110 R. Zapperi, *L'Homme enceint*, 'Les Chemins de l'Histoire' (Presses Universit-
    aires Françaises, Paris, 1983), p. 131.
111 *In IV Sententiarum*, distinction XXXI, G, article 24 (ed. A. Borgnet, p. 263).
    In the *De Animalibus*, Albert the Great refers solely to physical reasons.
    Certain positions are unfavourable to conception because they do not allow
    the semen to be adequately received in the womb:

    Quando autem in latere disponitur, vix contingit quia in latus gutturis
    semen proiciatur. Quando autem mulier virum supergreditur, matrix est
    revoluta: et ideo effunditur id quod est in ipsa. Quando autem stat mulier,
    extenditur matrix et constringitur os eius, ut non recipiat, et si recipit,
    effunditur propter extensionem. Posterius autem cognita mulier non recipit
    semen, nisi inter labia vulvae, quia spissitudo natium impedit veretri usque
    ad os matricis porrectionem. Praeterea matrix tunc est reversata et non in
    situ naturali, et ideo non facile semen accipit. [Stadler, (ed.) bk X, tr. 2,
    ch. 1.]

    We notice once more that Albert shows no hesitation in going into a
    detailed description.
112 Jacques Despars (fifteenth century) tells us of his experience as a doctor
    on this subject. 'Novi aliquos stultos qui quando coibant dimittebant
    feminam complere coitum et sponte proprium sperma retinebant ut ipsum
    sufficeret ad plures coitus (unde delectabantur pluribus vicibus et magis
    placebant huic cum qua coibant frequentia coitus), quibus inde accidit circa
    pudibunda serpigo corrodens que diu duravit corpus circuiens et nisi
    cauteriis sanari non potuit.' (*Expositio supra Librum Canonis Avicenne*, bk III,
    fen 20, tr. 1, ch.7).
113 See G. Beaujouan, 'Manuscrits médicaux du Moyen Age conservés en
    Espagne', *Mélanges de la Casa de Velazquez*, 8 (1972), p. 173.
114 These proofs were given by Dr P. Bohigas to M. Guy Beaujouan, who was
    kind enough to hand them on to us.
115 See Beaujouan, 'Manuscrits médicaux', loc. cit.
116 Here is an extract from the prologue: 'E yo quir parlar per aço en aquesta
    raho e complir e be declarar, per so que entenen tot hom quil vulle guardar,

e ques pusquen aprofitar dell tambe los fisichs e los cirurgians e moltes
daltres gents; e vull guardar que no age longues rahons, mas ques pusquen
aprofitar de la cura e de la obra mallor.'

117 The woman must have four black attributes: her hair, eyelashes, eyebrows
and eyes; four red attributes; her complexion, tongue, gums and lips; four
white attributes: her complexion, teeth, the white of her eyes, her thighs.
Four parts of her body must be narrow: her nostrils and ear-holes, her
mouth, her nipples, her feet; four must be slender: her eyebrows, nostrils,
lips and ribs; four must be big: her forehead, her eyes, her breasts, her
haunches; four must be round: her head, neck, arms and legs. Four parts
of her body must give off a perfume: her mouth, nostrils, armpits and
vagina. Note that apart from this last item, the sex organs do not appear.

118 'Item, la muller que li ve la volentat atart e no ha talent de foder, que li
fáçe lom. V. coses: el besar, el palpar, el pessigar, el estrenyer, el farir ab
les mans.' The sensitive parts of the body are: the face, hands, legs, throat,
breasts, stomach and navel.

119 We will quote one of the last descriptions; it calls for the use of an accessory:

Item, altra manera: que sacost la fembra a la paret e que leu la perna
esquerra que la pos sobre. I. banch o qualque cosa altre, e lom que li stia
fodendo per detras e tenent ell la perna que te al banch en la ma total
vegada, car mils sacostara a ella e fara sa obra mils.

Note that the Tao proposes twenty-six postures; see Jolan Chang, *The Tao
of Love and Sex; the Ancient Chinese Way to Ecstasy*, foreword and postscript
by J. Needham (Panther, London, 1977), pp. 56–8.

120 Perhaps this work inaugurates a long series of practical treatises, thanks to
which people may gain better control over their bodies. On this new
approach, see M.M. Fontaine, 'L'Art de nager. Conceptions humanistes et
réligieuses: leur influence sur le développement de la natation du XVe au
XVIIe siècle, *Proceedings of the IX International HISPA Congress* (Min. de
Qualidade de Vida, Lisbon, 1982), pp. 132–42.

121 See above, p. 13.

122 'Luxuria viri primam radicem habet in lumbis et libido mulieris in umbilico.
Unde Job "Virtus eius in lumbis eius et fortitudo eius in umbilico ventris
eius [XL, 15–17]"'; Michael Scot, *Liber Physionomie* (B. de Garaldis, Pavia,
1515), fol. 9 v. On the erotic role of this part of the body, see G. Tibon,
Le Nombril, centre érotique (Pierre Horay, Paris, 1983).

123 One of the first doctors to denounce this danger was Bernard of Gordon
(end of thirteenth to beginning of fourteenth century); see D. Jacquart, 'La
Réflexion médicale médiévale et l'apport arabe', *Nouvelle Histoire de la
psychiatrie* (Privat, Toulouse, 1983), p. 51.

124 *Expositio supra Librum Canonis Avicenne*, bk III, fen 20, tr. 1, ch. 12. The text
of Avicenna that is thus expanded was: 'Et oportet ut referantur ei proposita
coeuntium et libri narrati de dispositionibus coitus et figuris ipsius.'

Chapter 4   The Innocent and the Guilty

1 Thomasset, (ed.) *Placides et Timéo,*, p. 206, § 424.

2 *Quaestiones super De Animalibus*, Filthaut, tr. XVIII, Q 3, 'Utrum causa

assimilationis sit aliqua virtus in semine.'

3 MS Paris, Bibliothèque nationale, Latin 15456, fol. 188 r.

4 Nicaise, (ed.) *La Grande Chirurgie de Guy de Chauliac*, p. 547. Middle English tr. in Margaret S. Ogden, (ed.) *The Cyrurgie of Guy de Chauliac* (Oxford University Press, Oxford, 1971), p. 529.

5 See Thomasset, *Commentaire du Dialogue*, pp. 32 and 162.

6 Pseudo-Galen, *De Spermate*, MS cited, fol. 188 v.

7 The melancholic person is also characterized by a softness, a sloth in resisting evil; he or she is more inclined to the sin of acedia; see S.W. Jackson, 'Acedia the Sin and its Relationship to Sorrow and Melancholia in Medieval Times', *Bulletin of the History of Medicine*, 55 (1981), pp. 172–85.

8 Lawn, *The Prose Salernitan Questions*, B8, p. 6. The author of this question shares out these different aptitudes beween masters of Hereford, living at the turn of the thirteenth century.

9 Michael Scot, *Liber Physionomie* (B. de Garaldis, Pavia, 1515), 10 v.–11 r.

10 *Aucassin et Nicolette: chante-fable du treizième siècle, mise en français moderne par G. Michaut* (de Boccard, Paris, 1964), p. 49. Our English translation is taken from *Aucassin and Nicolette and Other Tales*, tr. Pauline Matarasso (Penguin, Harmondsworth, 1971), p. 35.

11 See the edition of the translation of Bartholomew of Messina: R. Förster, *Scriptores Physionomonici Graeci et Latini*, (G. Teubner, Leipzig, 1893), I, p. 39.

12 Rhazes, *Liber ad Almansorem*, II, 55.

13 Scot, *Liber Physionomie*, fol. 16 r.; Michael Savonarola, *Speculum phisionomie* MS Paris, Bibliothèque nationale, Latin 7357, fol. 30 v.

14 See A. Denieul-Cormier, 'La Très Ancienne Physiognomie et Michel Savonarole', *Biologie médicale*, 45 (1956).

15 *Quaestiones super De Animalibus*, ed. Filthaut, IX, Q 18.

16 See P. Duhem, *Le système du monde* (Hermann, Paris, 1913), vol. I, pp. 373–4. The reference is to the definition given by Simplicius.

17 *Quaestiones super De Animalibus*, tr. V, Q 10.

18 Galen, *Techne*, II, Avicenna, *Canon*, bk III, fen 20, tr. 1, ch. 4.

19 Note that the ancient description of satyriasis, which sometimes included a fatal prognosis, does not now correspond to any nosological entity. Certain serious cases have been interpreted as intoxications caused by an abuse of aphrodisiacs; see Schumann, *Sexualkunde und Sexualmedezin*, column 85.

20 We should point out that the ascription of this work to Arnald of Villanova is no longer generally accepted. See edn in *Opera Arnaldi de Villanova* (F. Fradin, Lyon, 1504), fol. 176 v.

21 *Expositiones in Librum Tertium Canonis Avicenne* fen 20, tr. 1, ch. 22.

22 Valescus of Tarentum. *Philonium* (S. de Gabiano, Lyon, 1535), fol. 326 v.

23 See M.D. Grmek, *Les Maladies à l'aube de la civilisation occidentale* (Payot, Paris, 1983), pp. 214–21.

24 This is the advice given by Bernard of Gordon, *Lilium Medicine*, part. VII, ch. 4, p. 604.

25 *Opera Arnaldi de Villanova*, fol. 100 v.

26 Ibid., fol. 101 r.

27 The reproach expressed by a seventeenth-century author in connection with masturbation cannot be applied to medieval doctors: 'Whence it comes about that doctors sin grievously, who advise this act for the health, and those who follow this advice are in no way exempt from mortal sin'; see Flandrin, *Le Sexe et l'Occident*, p. 263.

28 See P. Diepgen, *Die Theologie und der ärtzliche Stand* (Walter Rothschild, Berlin, 1922), p. 57.

29 *Lilium Medicine*, p. 604; *Philonium*, edn cited, fol. 327 r.

30 *Glosule super Viaticum*, MS Paris, Bibliothèque nationale, Latin 6891, fol. 57 v.

31 *De Bono*, ed. H. Kühle, *Alberti Magni Opera Omnia* (Aschendorff, Münster, 1951), vol. 28, pp. 160–3.

32 See *Opera Arnaldi de Villanova* (Lyon, 1504 edn) fol. 176 v.

33 *Livre de contemplacioun*, ch. 143, § 4; in the *Livre des mille proverbes*, one finds also (Proverb 7): 'The paths of lust are sight and hearing, its dwelling-place is imagination, its bed is will.' We base our discussion on the lecture given by Armand Llinarès, cited above, 'Raymond Lulle, un fou d'amour'.

34 See J. Pigeaud, 'Le Rêve érotique dans l'Antiquité grèco-romaine: l'oneirogmos', *Littérature, Médecine, Société*, 3 (1981), pp. 10–11. The translation we are quoting is taken from Caelius Aurelianus, *Chronic Diseases* V, 7, ed. and tr. I.E. Drabkin (University of Chicago Press, Chicago, 1950), pp. 958–63.

35 Note that the course of treatment for the body involves all the elements that we find in the Middle Ages being used to prevent involuntary emissions: a hard, cold bed, a sheet of lead placed on the small of the back (mentioned by Galen), cold food and cold baths, etc. See Pigeaud, 'Le Rêve Erotique', p. 11.

36 These comparisons can be found in the *Glosule super Viaticum* of Girardus Bituricensis and the *Philonium* of Valescus of Tarentum.

37 M. Foucault, 'Le Combat de la chasteté', *Communications*, 35 (1982), p. 21.

38 See edn G. Elsässer, 'Ausfall des Coitus als Krankheitsursache in der Medizin des Mittelalters', *Abhandlungen zur Geschichte der Medezin und der Naturwissenschaften*, 3 (1934), pp. 15–20.

39 *De Animalibus*, ed. Stadler, bk IX, tr. 1, ch. 1, § 7, p. 675.

40 He adds, as we shall see, homosexual practices.

41 *Opera Arnaldi de Villanova*, fol. 190 v.–191 r.

42 Ibid., bk III, ch. 6.

43 See above, p. 46.

44 See Albert the Great, *Quaestiones super De Animalibus*, ed. Filthaut, bk X, Q 5.

45 *Lilium Medicine*, p. 604.

46 MS cited, fol. 58.

47 *Philonium*, fol. 327 r.

48 Flandrin, *Le Sexe et l'Occident*, p. 133.

49 *Le Livre des bons usages en matière de mariage*, tr. L. Bercher (Maisonneuve et Larose, Paris, 1953), p. 91.

50 Māsawaih (Mesue), *Le Livre des axiomes médicaux*, pp. 196 and 197.

51 In the thirteenth century, the second translator, Giles of Santarem, put these aphorisms into Latin; ibid., p. 245.

52 'Just as some men enjoy the habit of eating other human beings, or of uniting with animals or other men, or other such activities, which are not in accordance with human nature' – *S. Thomas de Aquino Ordinis praedicatorum Summa Theologiae Cura et Studio Instituti Studiorum Mediaevalium Ottaviensis* (Ottawa, 1953), vol. II, col. 885b–886a (II–1, Q 31, article 7).

53 *Expositiones in Librum Tertium Canonis Avicenne*, fen 20, tr. 1, ch. 6.

54 *Canon*, bk III, fen 20, tr. 1, ch. 36.

55 *Expositiones*, passage from the *Canon* cited in the previous note.

56 See Aristotle, *Problems I–XXI*, tr. W.S. Hett (Loeb Classical Library, Cambridge, MA, and London, 1961), pp. 126–31. The author of the *Problems*

suggests a physical explanation for 'sexual passivity' which is taken up again in the fourteenth century by Peter of Abano in his commentary (cf. N.G. Siraisi, 'The *Expositio Problematum Aristotelis* of Peter of Abano', *Isis*, 6 (1970), p. 336). The fact that Gentile da Foligno neither develops nor discusses, nor even quotes properly, the Aristotelian position, clearly shows his refusal to tackle the problem from a medical point of view.

57 *Expositio supra Librum Canonis Avicenne*, bk III, fen 20, tr. 1, ch. 36.

58 Ibid., bk III, fen 20, tr. 1, ch. 6.

59 We will quote one of the questions asked under the rubric 'Adultery' in *Confessio Debet* by Hugh of Saint-Cher (mid-thirteenth century): '"Or have you sinned with your own wife against nature?" If the sinner inquires, "What's that – 'against nature'?" the priest may say, "The Lord gave one way which all men hold; wherefore, if you have done other than in that one way you have sinned morally."' For the problem as a whole, see Noonan, *Contraception*, pp. 273ff.

60 'Mulieres enim secundum plurimum tardantur in emittendo sperma et remanent non complentes desiderium suum, quare non fit generatio. Et iterum ipse remanent secundum desiderium suum quare ille qui ex ipsis non custodiuntur mittunt in illa dispositione super se ipsas quas inveniunt et propter hanc causam redeunt ad fricationem ut perveniat in eo quod est inter eas complementum voluntatis' – *Canon*, bk III, fen 20, tr. 1, ch. 38.

61 *Expositio*, reference to the *Canon* cited in the previous note.

62 J. Le Goff, *The Birth of Purgatory*, tr. Arthur Goldhammer (Scolar Press, London, 1984), p. 178.

63 Flandrin, *Le Sexe et l'Occident*, pp. 114–15.

64 Christianity, Social Tolerance and Homosexuality, pp. 316–18.

65 *In Evangelium Lucae*, XVII–29, ed. A. Borgnet, (L. Vivès, Paris, 1895), vol. 23, p. 488.

66 J. Boswell, *Christianity, Social Tolerance and Homosexuality* (Chicago University Press, Chicago, 1980), p. 317.

67 *De Animalibus*, ed. Stadler, XXII, 2, 1, § 122.

68 J.D. Rolleston, 'Penis Captivus: A Historical Note', *Janus*, 39 (1936), pp. 196–201. The definition is as follows: '"Penis captivus" is applied to incarceration of the organ in the vagina due to psychogenic spasmodic contraction of the levator ani, and not to the condition resulting from the insertion of the penis into rings and similar inanimate objects' (p. 196). Lucretius is considered to be the source: *De Rerum Natura IV*, 1197–1201. For the Middle Ages, the author quotes the poem by Robert de Brunne, alias Robert Mannying (1264–1340), a translation of the *Manuel des Pechiez* by William of Wadington.

69 Saxo Grammaticus, *Historia Daniae* 1644, lib. XIV 327–28.

70 *Livre du Chevalier de La Tour Landry pour l'enseignement de ses filles* (A. de Montaiglon, Paris, 1854), ch. 35. We are quoting from the English text by Caxton, 1484.

71 In this chapter on the fear of women, we would place the discussion of the 'penis captivus' (a very rare medical accident) alongside variations on the theme of the 'vagina dentata' dear to anthropologists.

72 Vincent of Beauvais, *Speculum Naturale* (D. Nicolinum, Venice, 1591), bk XIX, ch. 118.

73 *De Universo*, III, ch. 25; *Opera Omnia* (D. Zenari, Venice, 1591), p. 1009.

74 *Speculum Naturale* (Venice, 1591 edn), bk XXII, ch. 41.

75 *Generation of Animals*, tr. Peck 769b, p. 419.
76 Lawn, *The Prose Salernitan Questions*, no. 12, pp. 285–87. The formation of this monstrous creature may also refer to the runt mentioned by Aristotle in connection with pigs (*Generation of Animals*, 748b).
77 Lawn, *The Prose Salernitan Questions*, p. 286, note 12.
78 See J. Vernet, 'Los Médicos Andaluces en el *Libro de las Generaciones de Médicos*, – ibn Yûlyûl', *Anuario de Estudios Medievales*, 5 (1968), pp. 456–7.
79 *De Universo*, p. 1009.
80 *De Causa Primaria Paenitentiae in Hominibus et de Natura Daemonum*. We are using the edition by J. Burchardt, Studia Copernica, XIX (Polska Akademia Nauk, Wroclaw, Warsaw, Krakow and Gdansk, 1979), pp. 161–99. For the demonology of Vitelo, see also the works of Eugenia Paschetto, especially *Demoni e Prodigi. Note su alcuni Scritti di Witelo e di Oresme* (Turin, 1978).
81 *De Causa Primaria Paenitentiae*, ed. J. Burchardt, p. 173.
82 See above, page 85. Note that in 1637, the imagination was still held responsible for the conception of a child in a woman, four years after the departure of her husband. See P. Darmon, *Le Mythe de la procréation à l'âge baroque* (Seuil, Paris, 1981), p. 107.
83 *De Causa Primaria Paenitentiae*, ed. J. Burchardt, p. 177.
84 Caelius Aurelianus, *Chronic Diseases*, I, 3, ed. and tr. I.E. Drabkin, pp. 474–7.
85 Bernard of Gordon, *Lilium Medicine*, II, ch. 24, pp. 220–1.

Chapter 5   The Exposed Body

1 See J. Löffler. 'Die Störungen des geschlechtlichen Vermögens in der Literatur der autoritativen Theologie des Mittelalters', *Abhandlungen der Akademie der Wissenschaften und der Literatur im Mainz, Geistes- und Sozialwissenschaftlichen Klasse*, 6 (1958), pp. 296–380.
2 P. Darmon, *Le Tribunal de l'impuissance, virilité et défaillances conjugales dans l'Ancienne France* (Seuil, Paris, 1979).
3 *Pantegni, Practica*, bk VIII, chs 1, 2, 28 and 29; *Canon*, bk III, fen 20, tr. 1, chs 10, 11, 12, 29; fen 21, tr. 1, chs 7, 8, 9; tr. 4, chs 1, 2, 3.
4 'De Sterilitate tam ex Parte Viri quam ex Parte Mulieris'; L. Thorndike and P. Kibre, *A Catalogue of Incipits of Medieval Scientific Writings in Latin* (Mediaeval Academy of America, London, 1963), cols 662, 1377, 1379.
5 *Lilium Medicine*, part VII, ch. 1.
6 *Rosa Anglica* (Augsburg, 1595 edn).
7 On the surgical treatment of hernias, see Huard and Grmek, *Mille ans de chirurgie*, p. 70. Note that the testicle was 'very often sacrificed by peripatetic herniotomists'; professional surgeons, however, attempted to avoid this castration.
8 Bk VI, ch. 2, 'De Defectu Coitu'.
9 *Registre des causes civiles de l'officialité épiscopale de Paris 1384–1387*, published by Joseph Petit (Imprimerie nationale, Paris, 1919), cols 86 and 299, in the series 'Collection de documents inédits sur l'histoire de la France'.
10 Ibid., col. 86.
11 Edition of Guy of Chauliac's Latin text in G. Hoffman, 'Beiträge zur Lehre von der durch Zauber verursachten Krankheit und ihrer Behandlung in der Medizin des Mittelalters', *Janus*, 37 (1933), pp. 191–2.

12 Ibid., pp. 129–44, 179–92, 211–13.
13 D. Jacquart, 'Le Regard d'un médecin sur son temps: Jacques Despars (1380?–1458)', *Bibliothèque de l'Ecole des Chartes*, 138 (1980), p. 63.
14 Chapter ed. and tr. in H.E. Sigerist, 'Impotence as a Result of Witchcraft', in *Essays in Biology in Honour of Herbert M. Evans* (University of California Press, Berkeley, 1948), pp. 541–4.
15 We are correcting the reading *in tecto* of the edition cited above *in lecto*.
16 See Grmek, *Les Maladies à l'aube de la civilisation occidentale*, p. 321.
17 I. Veith, *Hysteria: the History of a Disease* (University of Chicago Press, Chicago, 1965), p. 1.
18 J. Corraze, 'La Question de l'hystérie', *Nouvelle Histoire de la psychiatrie* (Privat, Toulouse), p. 401.
19 *Timaeus*, 91C, p. 250.
20 See Vazquez Bujan, 'Escritos Médicos Latinos Tardios, Estudio y Edición Critica de la Antiqua Traducción Latina del Tratado *Peri Gynaikeion*' (unpublished dissertation for the Faculty of Philology of the University of Santiago de Compostela, 1981, pp. 242–4.
21 See Rousselle, *Porneia*, p. 95.
22 See edn in *Opera Ysaac* (B. Trot and J. de Platea, Lyon, 1515), bk VI, ch. 11.
23 See Veith, *Hysteria*, pp. 101–2.
24 Bk III, fen 21, tractate 4, chs 16, 17, 18.
25 See Rousselle, *Porneia*, p. 95.
26 See Veith, *Hysteria*, p. 106.
27 *Rosa Anglica*, pp. 595–6.
28 *Expositio supra Librum Canonis*, bk III, fen 21, tr. 4, ch. 17.
29 Latin text edited in Löffler, 'Die Störungen des geschlechtlichen Vermögens', p. 350.
30 P. Diepgen, *Frau und Frauenheilkunde in der Kultur des Mittelalters* (George Thieme Verlag, Stuttgart, 1963), pp. 168–71.
31 Grmek, *Les Maladies à l'aube de la civilisation occidentale*, pp. 199–225. Our chapter on veneral diseases owes a great deal to this work, and to our conversations with Dr Grmek. For a bibliography of the controversy between Americanists and non-Americanists, see: F. Guerra, 'The Dispute over Syphilis: Europe versus America', *Clio Medica*, 13 (1978), pp. 39–61.
32 C.J. Hackett, 'On the Origin of the Human Treponematoses', *Bulletin WHO*, 1963, pp. 7–41; 'The Human Treponematoses', in D.R. Brothwell and A.T. Sandison, *Diseases in Antiquity* (Thomas, Springfield, 1967), pp. 152–69.
33 Grmek, *Les Maladies*, p. 207.
34 Ibid., p. 209.
35 These too-hasty interpretations can be found in B.L. Gordon, *Medieval and Renaissance Medicine* (Peter Owen, London, 1959), pp. 524–38. Another allusion to *sahaphati* can be found in F. Guerra, 'The Description of Syphilis in Avicenna', *XXVII Congreso Internacional de Historia de la Medecina, Actas* (Acàdemia de Ciències Mèdiques de Catalunya i Balears, Barcelona, 1981), pp. 731–3.
36 Bk III, fen 7, tr. 3, chapters 1 and 2.
37 *Liber Theisir* (O. Scoti, Venice, 1530), bk I, tr. 1, ch. 7.
38 *Tractatus de Chirurgia* (edited after the *Philonium*), C 16.
39 *Liber Theisir*, bk II, tr. 4, ch. 2.
40 *Les Maladies*, p. 215.
41 See also our comments on cases of leucorrhoea, above, p. 68.

42 This description is taken from A. King and C. Nicol, *Venereal Diseases*, 4th edn, (Baillière Tindall, London, 1975), pp. 258–63. See also A. Wisdom, *A Colour Atlas of Venereology*, Wolfe Medical Atlases, 6 (Weert, London, 1973).

43 See H.M. Fay, *Lépreux et cagots du Sud-Ouest (H. Champion, Paris, 1910), pp. 31–32.*

44 *Grmek, Les Maladies, p. 223.*

45 See E. Wickersheimer, 'Textes médicaux chartrains des IX^c, X^c et XI.^{c} siècles', in *Science, Medicine and History, Essays Written in Honour of Charles Singer* (Oxford University Press, Oxford, 1953), vol. I, p. 170.

46 See J. Dubois, *Un sanctuaire monastique au Moyen Age: Saint-Fiacre-en-Brie* (Droz, Geneva, 1976), pp. 139 and 141–2, note 346.

47 Description taken from J. Delanoë and A. Puissant, 'Les Affections transmises par voie génitale', *Concours médical*, 1977, pp. 4896–7.

48 *Les Maladies*, p. 224.

49 *La Cirurgie de maistre Guillaume de Salicet*, tr. Nicole Prevost (M. Husz, Lyon, 1492), ch. 48.

50 Grmek, *Les Maladies*, p. 253.

51 On this point, see a frequently quoted work: S.N. Brody, *The Disease of the Soul: Leprosy in Medieval Literature* (Cornell University Press, Ithaca, NY, and London, 1974).

52 See E. Jeanselme, 'Comment l'Europe au Moyen Age se protégea contre la lèpre', *Bulletin de la Société d'histoire de la médecine*, 1931, pp. 33–41; A. Bourgeois, *Lépreux et maladreries du Pas-de-Calais (X³–XVIII^c siècle)* (Commission départementale des Monuments historiques du Pas-de-Calais, 1972). Fourteenth-century medical texts bear witness to the growing participation of doctors or surgeons on the examining juries. Thus, as we noted in regard to impotence, a real effort was made to classify signs, multiply tests and evaluate the risks of error. We should not infer from the lack of precision and the confusion of certain disciplines that doctors made light of diagnosing leprosy. The fact that the face was affected was recognized as an infallible sign, and so authors such as Bernard of Gordon or John of Gaddesden wondered, not without great anxiety, whether they should declare a patient whose face was intact to be leprous. Cf. L. Demaitre, 'The Description and Diagnosis of Leprosy by Fourteenth-Century Physicians', *Bulletin of the History of Medicine*, 59 (1985), pp. 327–44; F.O. Touati: 'Facies leprosorum: réflexions sur le diagnostic facial de la lèpre au Moyen Age', *Histoire des sciences médicales*, 20 (1986), pp. 57–66.

53 For an initial approach to the topic, see the summary in Gordon, *Medieval and Renaissance Medicine*, pp. 489–94.

54 Grmek, *Les Maladies*, p. 234.

55 Oribasius, *Collectio Medica*, XLV, 28, quoted by Grmek, ibid., p. 249.

56 P. Legendre, *L'Amour du Censeur* (Seuil, Paris, 1974), p. 150. The text quoted is taken from O. Pontal, *Les Statuts de Paris et le Synodal de l'Ouest (XIII^c siècle)*, Paris, Bibliothèque nationale, 1971, pp. 205–7.

57 *Lilium Medicine*, part I, p. 89.

58 Quoted by Gordon, *Medieval and Renaissance Medicine*, p. 534.

59 *Lilium Medicine*, loc. cit.

60 See Grmek, *Les Maladies*, p. 228.

61 H.M. Fay, *Lépreux et Cagots*.

62 William of Conches, *Dialogus de Substantiis*, pp. 243–4. See also Adelard of Bath, '*Quaestiones Naturales*', ed. M. Müller, *Beiträge zur Geschichte der Philosophie*

und der Theologie des Mittelalters, 31 (1934), question 41. At the beginning of the fourteenth century, Peter of Abano, in his *Conciliator*, proposed the same kind of explanation: because of the male heat, the pores in the virile member open and this allows the vapours from the corrupt womb to enter. On the other hand, because her womb is cold, dry and dense, the woman is less receptive and she will contract leprosy only if she has sexual intercourse with numerous leprous men (see H. Lemay, 'Masculinity and Femininity in Early Renaissance Treatises on Human Reproduction', *Clio Medica*, 18 (1983), pp. 28–9).

63 R. Chaussinand, *La Lèpre* (Crépin-Leblond, Paris, 1950). These statements rest on a statistical study based on the disease in Indo-China:

> Sex seems to be equally involved in the receptivity of the organism to the leprous infection. The majority of writers note that leprosy attacks men more frequently than women. Indeed, it can be observed that in most countries affected by leprosy, there is a clear preponderance of male sufferers. The most commonly noted proportion is that of two male lepers for every female leper. [p. 37]

The author puts forward the following type of argument: man is supposedly vulnerable during puberty, whereas woman seems to be particularly resistant during that period of her life.

64 J. Rossiaud, 'Prostitution, sexualité, société dans les villes françaises au XVe siècle', *Communications*, no. 35: *Sexualités occidentales*, (Seuil, Paris, 1982), pp. 68–83. The exclusion of prostitutes seems to have considerably weakened in the fourteenth and fifteenth centuries.

> The observation of the social interdicts to which prostitutes were subject seems also to have been extremely partial and frequently forgotten. The marks of infamy which made of the prostitute an *untouchable* who had to be recognized immediately so as to be avoided were hardly any longer in force. To be sure, in the statutes of Avignon of 1441, it was recorded that *meretrices* were expected to buy any food that they might, in the market, have touched with their hands; but these statutes are largely a repetition of those of the thirteenth century, and it may be doubted that they were rigorously applied, especially when it is known that in the neighbouring towns of Languedoc (Nîmes in particular), during the Major Charity (Ascension Day), prostitutes themselves made cakes which the consuls publicly accepted so as to give them to the poor.

Over and above the changing moral attitudes, these facts suggest both that the diseases we are discussing were in retreat, and that their incidence may have varied from one geographical area to another.

65 C. Thomasset, 'La Femme au Moyen Age. Les composantes fondamentales de sa représentation: immunité–impunité, *Ornicar*, 22–3 (1981), pp. 223–38. The author was attempting to examine some of the consequences of these beliefs for the psychology of medieval people. The consequences were indeed as stated, but the determination of the cause (treponematosis) must be revised in favour of lymphogranuloma venereum.

66 See above, pp. 74–75.

67 This text occurs in the *Dialogue de Placides et Timéo*. On its principal sources and some European versions of the legend, see Thomasset, *Commentaire du Dialogue*, pp. 71–110.

68 *De Animalibus*, VII, 2, 5, ed. Stadler, *Albertus Magnus, De Animalibus*, pp. 553–4.

69 D. Bois-Bruxelle, 'Cinq Miracles de la Vierge tirés de la Deuxième Vie des Pères' (unpublished dissertation, University of Paris–IV, 1983), pp. 224–98.

70 See Beroul, *The Romance of Tristan*, tr. Alan S. Fedrick (Penguin, Harmondsworth, 1970), pp. 132 and 136.

Conclusion

1 A. Burguière, 'De Malthus à Max Weber; le mariage tardif et l'esprit d'entreprise', *Annales ESC*, 1972, p. 1131.

2 Noonan, *Contraception*, p. 298.

3 Diepgen, *Die Theologie*, p. 9, note 30.

4 Ibid., p. 54, note 319.

Index

abortion 87, 88, 89, 91, 93, 113, 115, 164

abscesses 181, 183

abstinence 86, 117, 136, 147, 152

acedia 225

Adelard of Bath 189

adolescents 88, 93, 99, 218

adultery 64, 82, 85, 89, 99, 153, 166, 215

Afflacius, Johannes 214

al-Ghazali 154

Alan of Lille 105, 108

Albert 91, 93

Albert the Great 38–40, 44–7, 55–7, 66–9, 75, 79–82, 85, 90, 93, 111, 128, 129, 135, 139, 141, 146, 153, 160–1, 176, 192, 195

Albert of Trebizond 64

Albucasis 22, 45, 141, 207

alchemy 5, 59, 112

Alcmaeon of Crotona 53, 61

Alexandre of Hales 93

Alfanus of Salerno 22, 60

'Ali ibn al-'Abbas al-Majusi 22, 31, 45, 62, 71–2

allegory 14, 16, 99

amenorrhoea 73

amplexus reservatus 96, 195

analogy 5, 8, 10, 13–15, 25, 27, 32, 36–7, 46, 54, 59, 63, 72–4, 85, 91, 105, 119, 161

anaphrodisiacs 119

Anatomia Cophonis 26, 27, 32

Anatomia Magistri Nicolai Physici 15, 26, 32–3

Anatomia Vivorum 36–8

anatomy 5, 7–47, 52, 66, 79, 88, 132, 138, 170–1, 177

Anaxagoras 53, 61

animals 7, 26–7, 30–2, 40, 70–5, 100, 118, 123, 129, 132, 139, 161, 163–7, 191, 197

Anthony of Florence 149

antiperistasis 146

anus 24, 120, 124, 131, 182; anal practices 88, 107

aphrodisiacs 91–2, 114, 119, 225

Aphrodite 209

apoplexy 165, 175

appetite (sexual) 80, 119, 148, 170, 177

Aquinas, Thomas 57, 76–7, 93, 111, 155–6, 213–14

Aristotle 3, 5, 15–16, 22, 36–9, 54–69, 73–83, 90–1, 103, 110–11, 128, 130, 139, 141, 146, 163–4, 191, 194–5, 204; pseudo-Aristotle: *Physiognomy* 144–5, *Problems* 158

Arnald of Villanova 84–5, 118, 148, 170, 176, 225

arpo 164

arteries 30, 33, 41–3, 48–50, 60, 65, 71–2, 83, 118; aorta 30

as-Samau'al ibn Yahya 124

ascetic life 83